Global Environmental Change in Alpine Regions

NEW HORIZONS IN ENVIRONMENTAL ECONOMICS

This important series is designed to make a significant contribution to the development of the principles and practices of environmental economics. It includes both theoretical and empirical work. International in scope, it addresses issues of current and future concern in both East and West and in developed and developing countries.

The main purpose of the series is to create a forum for the publication of high quality work and to show how economic analysis can make a contribution to understanding and resolving the environmental problems confronting the world in the twenty-first century.

Recent titles in the series include:

International Climate Policy to Combat Global Warming
An Analysis of the Ancillary Benefits of Reducing Carbon Emissions
Dirk T.G. Rübbelke

Pollution, Property and Prices
An Essay in Policy-making & Economics
J.H. Dales

The Contingent Valuation of Natural Parks
Assessing the Warmglow Propensity Factor
Paulo A.L.D. Nunes

Environmental Policy Making in Economics
with Prior Tax Distortions
Edited by Lawrence H. Goulder

Recent Advances in Environmental Economics
Edited by John A. List and Aart de Zeeuw

Sustainability and Endogenous Growth
Karen Pittel

The Economic Valuation of the Environment and Public Policy
A Hedonic Approach
Noboru Hidano

Global Climate Change
The Science, Economics and Politics
James M. Griffin

Global Environmental Change in Alpine Regions
Recognition, Impact, Adaptation and Mitigation
Edited by Karl W. Steininger and Hannelore Weck-Hannemann

Environmental Management and the Competitiveness of Nature-Based
Tourism Destinations
Twan Huybers and Jeff Bennett

The International Yearbook of Environmental and Resource Economics 2003/2004
A Survey of Current Issues
Edited by Henk Folmer and Tom Tietenberg

Global Environmental Change in Alpine Regions

Recognition, Impact, Adaptation and Mitigation

Edited for the Human Dimensions Programme Austria (HDP-A) by

Karl W. Steininger
Professor at the Department of Economics
University of Graz, Austria

and

Hannelore Weck-Hannemann
Professor of Political Economy
University of Innsbruck, Austria

NEW HORIZONS IN ENVIRONMENTAL ECONOMICS

Edward Elgar
Cheltenham, UK • Northampton, MA, USA

Published by
Edward Elgar Publishing Limited
Glensanda House
Montpellier Parade
Cheltenham
Glos GL50 1UA
UK

Edward Elgar Publishing, Inc.
136 West Street
Suite 202
Northampton
Massachusetts 01060
USA

A catalogue record for this book
is available from the British Library

Library of Congress Cataloguing in Publication Data

Global environmental change in alpine regions : recognition, impact, adaptation and mitigation / edited for the Human Dimensions Programme Austria (HDP-A) by Karl W. Steininger and Hannelore Weck-Hannemann.
p. cm.— (New horizons in environmental economics)
1. Climatic changes —Environmental aspects—Alps. 2. Climatic changes—Economic aspects—Alps. 3. Alps—Environmental condtions. I. Steininger, Karl W. II. Weck-Hannemann, Hannelore. III. Series.

QC989.A43 G56 2002
363.738`747'09436—dc21

2002028587

ISBN 1 84376 183 1

Printed and bound in Great Britain by MPG Books Ltd, Bodmin, Cornwall

Contents

PART II IMPACT

PART III ADAPTATION

Figures

Tables

Contributors

Ammann, Walter J. Director of the Swiss Federal Institute for Snow and Avalanche Research (SLF), Davos and Director of the Research Department 'Natural Hazards' at the Federal Research Institute for Forests, Snow and Landscape (WSL), Birmensdorf. His research focuses on integral risk management of natural hazards, snow and avalanche mechanics, climate change and winter tourism. ammann@slf.ch

Chateau, Bertrand Chairman of Enerdata s.a. (Grenoble, France), a private company concerned with world energy statistics and information. Research work at Enerdata focuses on long term studies and modelling of energy-transport-environment systems. bertrand.chateau@enerdata.fr

Friedl, Birgit Researcher and Lecturer at the Department of Economics, University of Klagenfurt, Austria. Earlier she managed the Human Dimensions Programme Austria. Her research is concerned with modelling sustainable resource management and environmentally sound transport. birgit.friedl@uni-klu.ac.at

Frondaroli, Alberto Director of Branch Office of Rome of Centro Studi sui Sistemi di Trasporto (CSST) SpA. His research focuses on transport modelling, transport forecasting and environmental impact assessment. a.frondaroli@csstrm.it

Keller, Mario Member of the Management Board of Infras (Zurich/Berne, Switzerland), a private consulting group. His work especially focuses on transport, environment and spatial planning, development and application of transport models, energy and emissions scenarios, and technology impact assessments. mario.keller@infras.ch

Kleissner, Anna Junior Researcher at the Institute for Advanced Studies and Scientific Research Carinthia, Klagenfurt, Austria, with research interests in regional economics, banking, energy economics and national accounting. kleissner@carinthia.ihs.ac.at

Kroemker, Doerthe Researcher at the Centre for Environmental Systems Research at the University of Kassel, Germany. Previously she worked for the German Advisory Council on Global Change (WBGU). In her research she focuses on agent-based concepts of vulnerability, environmental action theories and images of nature in an intercultural perspective. kroemker@usf.uni-kassel.de

Loster, Thomas Geographer and member of the GeoRisksResearch Department at the Munich Reinsurance Company, Munich, Germany. He is the head of the weather perils and climate section and also responsible for the statistical analyses of worldwide natural catastrophes and trend analyses. tloster@munichre.com

Madlener, Reinhard Senior Energy Economist at the Centre for Energy Policy and Economics (CEPE), Swiss Federal Institutes of Technology (ETHZ, EPFL, PSI), Zurich and former Managing Director of the Institute for Advanced Studies and Scientific Research Carinthia in Klagenfurt, Austria. His main research interests lie in energy economics (particularly renewable energies), energy policy design, diffusion of innovative energy technologies, sustainable development, and applied econometrics. madlener@cepe.mavt.ethz.ch

Meissl, Gertraud Research Assistant at the Institute of Geography, University of Innsbruck, Austria. Her research is especially concerned with natural hazards and geographical information systems. gertraud.meissl@uibk.ac.at

Molitor, Romain Co-Director at *Trafico* (Transport planning and consulting), Vienna, Austria. His research focuses on environmental and societal assessment of transport, transport planning approaches and urbanism and on transport modelling. romain.molitor@trafico.at

Mosler, Hans-Joachim Senior Researcher at the Department of Psychology, University of Zurich, Switzerland. On completing his post-doctoral thesis he became fully qualified as university Professor in psychology at the University of Zurich (habilitation). He has strong research interests in environmental psychology and environmental intervention in communities. mosler@sozpsy.unizh.ch

Payer, Martin Research Assistant at the Department of Economics, University of Graz, Austria. Before he worked at the Institute for Environmental Management and Business Administration at the European Business School, where he was concerned with the linkage between corporate environmental performance and shareholder value. martin.payer@uni-graz.at

Ploner, Alexander Consultant engineer at i.n.n. ingenieurbuero fuer naturraum-analyse und naturgefahren-management, ploner & soenser OEG innsbruck, Austria. office@inn-ingenieurbuero.at

Psenner, Roland Professor of Limnology and Head of the related research unit at the University of Innsbruck, Austria. His special research interests are in alpine lakes, especially in climatic warming effects on lakes and catchments, the cycling of elements and pollutants, and microbial ecology of high altitude lakes, ice and snow. roland.psenner@uibk.ac.at

Schleicher, Stefan P. Professor of Economics at the University of Graz, consultant to the Austrian Institute of Economic Research, has been involved in many political advisory bodies such as the Austrian Environmental Plan and the Austrian Council on Climate Change. His research focuses on concepts and models for sustainable development, in particular energy, climate and econometric issues.
stefan.schleicher@wifo.at.

Soenser, Thomas Consultant geologist at i.n.n. ingenieurbuero fuer naturraum-analyse und naturgefahren-management, ploner & soenser OEG innsbruck, Austria. office@inn-ingenieurbuero.at

Steininger, Karl W. Associate Professor of Economics at the University of Graz, Austria. Human Dimensions representative in the Austrian National Committee on Global Change. Previously he was with the World Bank. His research focuses on economic modelling, particularly in international and environmental economics, and on sustainability, transport and water issues. karl.steininger@uni-graz.at

Steininger, Klaus C. Senior researcher at the landscape planning consultancy *Raumumwelt*, Vienna, Austria. He is engaged in spatial planning at the local political management level. His research focuses on agriculture, environmental protection management and regional planning. steininger@raumumwelt.at

Stoeckli, Veronika Leader of the department 'Alpine Environment' at the Swiss Federal Institute for Snow and Avalanche Research, SLF, Davos. She has a research interest primarily in the vulnerability of ecosystems to natural hazards as well as in the ecology of the cryosphere. v.stoeckli@slf.ch

Stoetter, Johann Professor of Geography at the University of Innsbruck, Austria, with a research focus on natural hazards and environmental change. hans.stoetter@uibk.ac.at

Sulzer, Wolfgang Assistant Professor at the Institute of Geography and Regional Sciences, University of Graz, Austria. In his research he is especially concerned with aspects of high mountain geography (environmental change detection, climate and glacier research) and remote sensing. wolfgang.sulzer@uni-graz.at

Weck-Hannemann, Hannelore Professor of Political Economy at the Institute of Public Economics, University of Innsbruck, Austria. Prior to this she held a professorship at the Humboldt-University Berlin. She has strong research interests in public finance and environmental economics. hannelore.weck@uibk.ac.at

Weiss, Gerhard Research assistant at the Institute of Forest Sector Policy and Economics, University of Agricultural and Life Sciences, Vienna, Austria. His special field of interest lies in sustainable forest management policies, mountain forestry and protection against natural hazards. weiss@mail.boku.ac.at

Wohlgemuth, Norbert Associate Professor of Economics at the University of Klagenfurt, Austria. Previously he worked as Senior Economist at the UNEP Collaborating Centre on Energy and the Environment in Denmark and at the Economic Analysis Division of the International Energy Agency (IEA-OECD) in Paris. His research areas are in the field of energy economics. norbert.wohlgemuth@uni-klu.ac.at

Preface

Glacier retreat is significant. For example, extrapolating the trend in volume loss of the last 150 years indicates that only 25% of the glacier volume existing in 1850 might be left by 2030 in the Swiss Alps. This has a direct bearing on the availability and timing of freshwater to the surrounding lowlands. More generally, global warming as well as global environmental change are expected to have further significant impacts on sectors such as tourism, agriculture, energy production, transportation, and natural hazard management. The observed trends in mountainous regions led the authors contributing to this volume and many of their peers to come together and explore the following:

- Which observations have been confirmed by sound scientific analysis?
- Which impacts are already observable and which need further consideration?
- How did adaptation work in the past and what options for improvement are available?
- How can processes of global environmental change be mitigated?

The wide ranging nature of these questions is reflected not only in the broad spectrum of interrelated disciplines from which the team of authors is drawn, but also by the interdisciplinary character of the individual chapters of this book. After having made significant contributions within their own areas of research the authors agreed to pool their resources and engage in a common effort. Co-operation continued for some two years and culminated so far in the production of this volume. Cross-disciplinary research might take more time, and related to one area of present concern, it is likely that during the preparation of this book the surface temperature in the European Alps has risen by more then a tenth of a degree (considering mountainous areas, the warming, at least of the European Alps, seems to be occurring faster than world-wide averages indicate). Nevertheless, a thorough understanding of the workings of environmental and social processes is crucial for the development of suitable action plans.

Our special thanks go to Birgit Friedl and Martin Payer, who, within the framework of the Human Dimensions Programme Austria, have contributed most significantly to the enormous editing work across authors, disciplines and software, as well as to the general communication and organisation, in particular by arranging a workshop for authors to meet and discuss their ideas in person. Our thanks also go to the University of Innsbruck for hosting the workshop, and to the further participants of the workshop for their active involvement, presentations and contributions to discussions, Thomas Geist, Georg Grabherr, Michael Kuhn, Magdalena Margreiter, and Engelbert Theurl. This volume has strongly benefited from the cross-reviewing of its authors, and from four outside reviewers, Karst van Geurs, Henk Folmer, Wallace Oates, and Franz Sinabell. The Austrian Federal Ministry for Education, Science and Culture supported the international co-ordination necessary for this volume. The research leading to the individual chapters of this volume was in many cases (co-)funded and/or co-ordinated by national and international institutions (such as the OECD). These are noted in each of the respective chapters. They are thankfully acknowledged here from the overall perspective of interdisciplinary co-operation. The Austrian Network for Environmental research facilitated this interdisciplinary co-operation. We also thank Munich Re for the supply of material on natural hazards to the workshop participants. Laurie Conway has done an extensive job on improving the clarity of the English. Finally, Martin Payer thoroughly prepared the index references. Throughout the process our efforts were well supported and encouraged by the staff of Edward Elgar, most importantly in this case, by Dymphna Evans, Martin Pitman, Alison Stone, and Laura Wyer.

We sought to bring together a volume incorporating the most significant cutting edge research on the human dimensions of global environmental change in alpine regions. Together with all the contributing authors we are pleased to be able to pass on this 'mountain' of research results and ideas. May the paths explored here be of benefit to all those interested and active in the field.

Graz and Innsbruck, August 2002 K.W. St., H. W.-H.

1. Introduction and Overview

Karl W. Steininger and Hannelore Weck-Hannemann

Our planet, the supplier of humankind's livelihood, is being subjected to substantial global environmental change. Without doubt this has been one of the most crucial insights of the last decade. Starting in the late 1980s, with the gradual acceptance of climate change as a real phenomenon, general awareness has since expanded to include a broader range of anthropogenic changes. These affect land surfaces, the atmosphere, biological diversity, the water cycle or biogeochemical cycles clearly beyond natural variability.

What is the impact of such developments on humankind, more specifically on the physical environment, social institutions, economic activities and the well-being of individuals? Are there any reasonable adaptation options and will they be taken? Are there reasons for mitigating global environmental change and, if so, what are the options? What does each of the mitigation options promise? What would be the preconditions, instruments, and consequences of these options? Such is the range of questions that this book is devoted to.

While the aspects of environmental change mentioned above are a global phenomenon, their impact tends to be of a more local (albeit uneven) nature, and thus options for adaptation and mitigation can vary widely across regions. This is the underlying motive behind the choice of a specific region as the basis for analysis in this book, and it should in fact aid the attainment of a more down to earth approach. People living in alpine regions are among those affected earliest and most significantly by global environmental change. On the one hand, indications of global environmental change can be found earlier in alpine environments than in areas of moderate climate (for example, mountain lakes are more sensitive to and better indicators for global environmental change than lakes in more moderate climates). On the other hand the consequences of such changes are more significant, as alpine populations do not have many options available when faced with the need to adapt to global environmental change. According to avalanche protection plans, for example, only a low proportion of land is suitable for housing. In addition, the population depends on relatively few economic sectors, foremost on tourism and to a lesser extent agriculture, each of which are highly dependent on the 'state of the environment'. Alpine populations are therefore particularly vulnerable.

While 'alpine' often refers to the European Alps specifically, the term has also been adopted to describe a well defined ecosystem situated above the timberline, and present around the world in all mountain ranges of sufficient height, from the Latin American Andes to the European Pyrenees.

In this introduction we first discuss the meaning and extent of global environmental change. A closer consideration of the term 'alpine' and what it entails then follows, before, finally, a detailed overview of the structure of the book and contents of individual chapters is given.

1 GLOBAL ENVIRONMENTAL CHANGE

What do we already know about global environmental change? We state some exemplifying crucial issues here.[1]

- In a few generations, humankind has embarked upon the process of exhausting fossil fuel reserves that were generated over several hundred million years. For example, the carbon dioxide concentration in the atmosphere has increased by more than 30 per cent since the beginning of the Industrial Revolution and methane has increased by 100 per cent.
- Nearly half the land surface has been transformed by direct human action, with significant consequences for biodiversity, nutrient cycling, soil structure and biology, and climate. More than one-fifth of terrestrial ecosystems have been converted into permanent croplands; most of the temperate, old growth forest has been cut. For example, more nitrogen is now fixed synthetically and applied as fertilisers in agriculture than is fixed naturally in all terrestrial ecosystems.
- More than 50 per cent of all accessible freshwater is used directly or indirectly by humankind; our underground water resources are being depleted rapidly. For example, dam construction alone has increased the volume of water held in river courses by 700 per cent over the last 50 years.

> Humans are altering the ecology of the planet, the chemistry of the planet, and the climate of the planet; moreover, these changes are coupled in an exceedingly complex manner. Multiple, concurrent interactions among the biota, between the biota and their environment, and between the environment and our social institutions may produce unpredictable results with cascading feedbacks.[2]

At the first large Global Change Conference where all international Global Change research programmes made contributions, a declaration was passed, stating inter alia that:

> Human activities are significantly influencing Earth's environment in many ways in addition to greenhouse gas emissions and climate change. [...] Global change is real and is happening now.[3]

To date, research initiatives on global change have to a large extent been natural science-based projects, involving relatively little input from social scientists. Many claim that the social sciences have not been equal partners in formulating the questions of global change research. In response to this lack of research, the socioeconomic dimension of global environmental change has recently gained momentum at the international level, and has received crucial support from the newly created International Human Dimensions Programme (IHDP) and its science projects. This book, too, focuses mostly on the social science dimension of global change, going beyond the pure natural science analysis, but still interacting with natural sciences. This extension of the research on global environmental change to socioeconomic analysis is considered crucial in order to be able to develop strategies of humankind to answer the global change challenge.

2 ALPINE REGIONS

In this book we argue that the impact of global environmental change has specific local and regional forms. We analyse alpine regions in particular, as these are more strongly exposed to global environmental change than many other regions. The issues of recognition, impact, adaptation and mitigation of global environmental change all need to be discussed in the specific setting of alpine environments.

The term 'Alpine' was first used to describe the geographical region of the European Alps (and in this case 'Alpine' is used with an initial capital letter), which is one focus of this book. However, alpine environments can be found all over the world. The term also describes ecosystems of a specific type. Alpine environments are situated above the timberline which means ca. 2 000–3 000 m altitude in temperate areas, about 4 000–5 000 m in tropical, and 0–1 000 m in boreal or polar regions.

Chapter 2 explicitly focuses on alpine environments in the latter (i.e. worldwide) sense. Other chapters are general and cover both the geographical and ecosystem definition (Chapters 3 and 6). The majority of chapters, however, focus on the geographical region of the European Alps in their specific empirical analysis, with many of the conclusions applicable worldwide to alpine environments or even more generally to mountainous environments.

The European Alps describe a mountain arc of about 800 km in length and 200 km in width from the Meditarranean Sea in the southwest to the edge of the central European plain on the northeast, with an average height of about 2 500 m a.s.l. and a maximum of more than 4 800 m a.s.l (Mont Blanc).

The European Alps may serve as an example for alpine environments, which because of their environmentally exposed location, were already the origin of a range of social and technological innovations in the past:

> The Alpine region has had a very dynamic history. Continual confrontation with the vagaries of an unpredictable climate coupled with the vulnerability of both man and animals in an area where natural catastrophes are frequent and often severe has traditionally stimulated innovative action. Seven hundred years ago, one of the first democratic systems emerged in the mountainous central cantons of Switzerland, triggered by the need for wise self-organisation by people living under extreme conditions and granted by the then-powerful European monarchs to guarantee free access to the important transit routes across the Alps. Alpine streams and rivers, fed during the whole year by snow and ice in the high mountains, provided an important course of mechanical energy. In those valleys, following the example of Great Britain, the industrial revolution first took hold on the European continent. Even now, the Alpine region is still responsible for many technical innovations. Its relative wealth and independence serves as a lifestyle model for much of the less developed world.[4]

3 STRUCTURE AND OVERVIEW

In this book solutions to a fourfold question are sought. These must be dealt with with respect to the particular exposure of alpine regions:

1. How can we understand the complex interlinkages within global environmental change and between global change and social and economic systems? (Recognition)
2. Of which type and magnitude are the most significant impacts of global environmental change on human societies? (Impact)
3. How does and can adaptation and adjustment to global environmental change take place? (Adaptation)
4. What are the options for mitigation in the sectors most significantly contributing to global environmental change in Alpine environments: agriculture, energy and transport? (Mitigation)[5]

3.1 Recognition

Alpine lakes represent the most remote and untouched aquatic environments in Europe, yet they are threatened by acid deposition, toxic air pollutants, increasing UV radiation, and climate change. Because of their sensitivity, remote mountain lakes are excellent sensors of change. In Chapter 2 Roland Psenner thus employs Alpine lakes as indicators of global change. They can be used to infer speed, direction and magnitude of impacts caused by changing air quality and climate trends. Natural processes interact with anthropogenic disturbances at different spatial and temporal scales, creating a complex pattern of links, dependencies, antagonisms and feedbacks. Nonetheless, Psenner shows that most interactions can be reduced to two chains of events resulting in either acidic lakes of low productivity, transparent to UV radiation and containing high amounts of metals, or in productive and alkaline waters protected from UV radiation and metal toxicity.

Our understanding of the interlinkage of global environmental change and social systems is also shaped by the scientific framework of analysis we have at hand. This is highlighted with a focus on climate change by Stefan Schleicher in Chapter 3. He begins by describing changes in the natural environment. The Alpine environment is likely to be affected by a global climate change not only through the projected increase of average temperatures but also through changes in many other climatic elements. In particular, changes in the magnitude and timing of precipitation would have significant impacts not only on the economy and ecology of Alpine regions but would also be felt far beyond the mountainous regions because of changes in the water cycle.

In many respects these events pose a challenge to the social science of economics. Conventional economic analysis would evaluate the impact on tourism and agriculture, the costs of relocating homes and production facilities, and the costs of adaptation. Identifying the deficiencies of this approach, Schleicher presents a comprehensive modelling framework that allows us to evaluate the impact of various measures of welfare, ranging from traditional production-based indicators to measures of sustainability and to restrictions based on cultural identity. The chapter thus supplies the theoretical foundations for evaluating and quantifying the social and economic impacts of global environmental change.

3.2 Impact

Snow cover is a crucial element in shaping Alpine environments. The changes in snow cover triggered by global change and their impacts on a broad range of consequences are analysed by Walter Ammann and Veronika Stoeckli in Chapter 4. In the European Alps, for example, snow, glaciers and permafrost dominate the area above 2 500 m above sea level. Between 1 500 and 2 500 m, the snow cover still predominates over vegetation and soil for more than half of the year, and is thus an important ecological factor, interacting with the flora, fauna and soil system. As an economic factor, the snow cover is most important in providing for a prosperous winter tourism industry. Snow also influences the safety on roads and railways and is an important water resource. A rise in temperature due to climate change will raise the temporary snow line, with subsequent impact on snow sports and thus on the economic situation of winter sport resorts at lower altitudes. Artificial snow may only partially be an appropriate and accepted alternative. Currently only very vague indicators for expected precipitation and weather extremes as a result of climate change are available. Based on these, the authors provide rough estimates of the economic effects of climate change. However, current indicators are not suitable for developing detailed economic impact studies. An increase in temperature will further raise the equilibrium line altitude of the glaciers with consequences for the annual water balance, and will probably cause permafrost terrain to thaw, potentially leading to instabilities of slopes. This causes impacts on mobility, hydroelectric power production, and agriculture and forestry, which the authors explore in detail. Besides changes in the abiotic environment, related changes in plant growth and distribution may also occur. Each of these changes may involve economic feedback as well.

One sector particularly exposed to the likely increase in extreme weather events and the increase in natural hazards is the insurance industry. Thus, in Chapter 5 Thomas Loster and Martin Payer elaborate on the link between 'mankind's gigantic experiment with the earth's climate' and its consequences for the insurance industry. An overview of the disaster trends in the last decades and their claims burden based on data from a major reinsurance company opens the contribution. On the basis of the Third Assessment Report of the Intergovernmental Panel on Climate Change, the most evident signs of climate change and possible scenarios for the 21st century are stated. The article also analyses ways in which the insurance industry tries to protect itself efficiently against future claims' burdens due to global change by integrating environmental aspects in its asset management and by adopting risk-sharing strategies with its clients and governmental institutions (e.g. the insured will be required to adopt a loss prevention behaviour or to bear a substantial deductible, and the insurance industry will continue to offer its risk management know-how to those concerned). This produces positive effects for both the insurance industry and the global environment.

3.3 Adaptation

This part of the book opens with a general view on why societies remain exposed to increasing risk (Chapter 6). Doerthe Kroemker and Hans-Joachim Mosler analyse the concept of vulnerability. The authors propose considering vulnerability as the absence of a person's or group's capacity to develop and implement protection and prevention measures. The chapter points out that capacity is influenced by two factors: the motivation and the competence to look for protection measures. A general framework, based on psychological action theory, is introduced that considers further psychological factors such as social norms, values, aims and knowledge, and explains why persons or groups have different motivations and competences and consequently different capacities for dealing with environmental criticality.

The specific aspects of the impact of global change processes on the natural hazard situation in Alpine countries are dealt with by Johann Stoetter, Gertraud Meissl, Alexander Ploner and Thomas Soenser in Chapter 7. Extrapolating from the developments of the recent past, they find that specific natural hazard processes are likely to occur more often than before and with higher intensity. Additionally, in most of the affected areas the damage potential has increased considerably, thus causing a higher risk potential. As at the same time public financial resources are becoming more and more limited, new ways of dealing with natural hazards have to be developed. The authors state which demands must be fulfilled in order to provide sustainable protection of the Alpine *lebensraum* and see natural hazard management as a promising concept. The planned 'alpS – Centre of Natural Hazard Management', which since the writing of this chapter, has actually been chosen for implementation by the federal research board, is discussed as an example.

Given the above, how does the State administration react to climate change and increased natural hazards? In Chapter 8 Gerhard Weiss tries to answer this question on the basis of the present political practice of natural hazards prevention in Austria. The article summarises the main results of an extensive research project on the application of policy instruments for torrent and avalanche control (e.g. mountain forest restoration, hazard zone mapping, subsidies for protective structures in torrent and avalanche control). The evaluation shows that although the official goal of natural hazards prevention is to produce sustainable, integrated and proactive hazards control, the relevant policies are only poorly implemented at present. The principles of sustainable hazards control are being undermined by short-term economic interests (e.g. extension of land utilisation). In practice, risk reductions are being outweighed by the increase in human activity in mountain areas. This trend implies that the present system's capability for precautionary action is limited, which is particularly critical considering the growing threats involved in climate change. The crucial point here involves not so much raising awareness within the public administration, but rather designing effective risk-prevention strategies. Presently, prevention goals do exist, but the institutions are not strong enough to implement them against powerful interests. Consequently, institutional changes would be necessary to redirect natural hazards policies. Concrete measures are proposed for a sustainable and more effective risk regulation in Austria. These include: stopping land development in areas situated below protective barriers; the reconsideration of levels of acceptable risk in hazard zone plans combined with catastrophe plans; reform of avalanche committees; better specification of subsidies combined with financial contributions from beneficiaries. In the case of climate change, all hazard zone plans would have to be revised and protective measures re-examined.

For adapting to the full range of global change impacts, Wolfgang Sulzer explores the instrument of climate suitability maps in regional planning (Chapter 9). Selecting the Austrian province of Styria as an illustrative example, a description of a characteristic climate in Alpine residential areas and its alteration by human activities is given. The article then describes the construction of three different map layers. The basic topographical map shows the different climatopes of the area of reference. The 'Cover Map I' includes information about the wind system between ground level and 50–100 m, the 'Cover Map II' sums up details about assessing building suitability in particular for land use planning.

3.4 Mitigation

While adaptation, as described in Chapters 6 to 9, aims at impact moderation, we can also, and most likely simultaneously, consider the sources of change more directly and seek to reduce their relevance. In Alpine environments one of the most important economic sectors is agriculture. Can a change in agricultural activities contribute to a reduction of climate change? Klaus and Karl Steininger seek to answer this question in Chapter 10. They find that the types of land use and the type and level of animal farming are the most crucial elements in such a reduction strategy. While there is some potential for mitigation, compared to annual greenhouse gas emissions from the whole economy, this potential is found to be relatively small, basically because agricultural management practices govern carbon stocks more than carbon flows.

The second sector where a major mitigation strategy could be implemented is energy production. How a gradual shift from fossil to renewable energy use can support a more sustainable development in Alpine environments, and particularly help to mitigate climate change and serve as a model for other regions, is discussed by Reinhard Madlener, Anna Kleissner and Norbert Wohlgemuth in Chapter 11. Locally available renewable energy sources often secure regional employment, and renewable energy technologies can often be developed, produced, and transferred in a regional context. Furthermore, local production and value added decreases the dependency on imports and thus benefits the regional economy and the security of supply. Social issues are enhanced as well: job security, better working and living conditions, and an increased participation of stakeholders in local and regional decision making are just a few examples analysed in this chapter. The Austrian province of Carinthia is taken as an illustrative example for the discussion of technical potential, actual market penetration, and CO_2 mitigation impacts of the various renewable energy technologies available.

Finally, transport is the sector with the highest increases in both greenhouse gas emissions and a range of other air pollutant emissions. In alpine environments the impacts of transport are especially severe: the European Alps in particular, are an outstanding example of a region where ecosystems and the environment are very vulnerable. This vulnerability is partly due to the topographical and climatic situation in the Alps. The shape of the valleys and meteorological particularities impede the escape of emissions from combustion processes. Hence, air pollution along these valleys is often as bad as in urban areas. Looking at sources, transport is among the major contributors in the Alpine region to environmental impacts of global and local scale.

A common research effort of Austria, France, Italy and Switzerland aiming at exploring the options for and the economic and social impacts of putting Alpine transport on an 'environmentally sustainable path' has been undertaken under the umbrella of the OECD. Bertrand Chateau, Birgit Friedl, Alberto Frondaroli, Mario Keller, Romain Molitor and Karl Steininger report on the results of this research in Chapter 12. These results indicate that a combination of technological measures and transport demand management can achieve tremendous reductions in environmental impact of Alpine transport. For example, the CO_2 emission objective of a reduction by 80 per cent relative to 1990 by the year 2030 (solely in transport) is proven feasible. The long-term aggregate economic impacts of such a strategy are minor, and, while significant restructuring and short-term costs occur, net social impacts have been found to be positive.

3.5 Conclusions and Future Research Prospects

Chapter 13 ends the book with a summary of the major conclusions of the above chapters and an outline of potential major research areas for the future.

NOTES

1. The facts cited are taken from Berrien Moore III, 'Challenges of a Changing Earth', Plenary presentation to the Open Science Conference Challenges of a Changing Earth, Amsterdam, 10 July 2001.
2. Moore (2001).
3. Amsterdam Declaration, published at the Global Change Open Science Conference Challenges of a Changing Earth, Amsterdam, July 2001.
4. Imboden (1998: 12–13).
5. While each of the sectors mentioned is also subject to adaptation to global change, the chapters in the part mitigation only focus on the potential of reducing the global change impact of these sectors' activities (mainly by reducing emissions).

BIBLIOGRAPHY

Imboden, D. (1998), 'Introduction', in Cebon, P., U. Dahinden, H. Davies, D.M. Imboden and C.C. Jaeger, *Views from the Alps: Regional Perspectives on Climate Change,* Cambridge, Massachusetts: MIT Press, 1–20.

Moore III, B. (2001), 'Challenges of a Changing Earth', Plenary presentation to the Open Science Conference, *Challenges of a Changing Earth*, Amsterdam, 10 July.

PART I

Recognition

2. Alpine Waters in the Interplay of Global Change: Complex Links – Simple Effects?

Roland Psenner

1 INTRODUCTION

While in the 1980s many scientific projects were focused on a better understanding and, eventually, the prevention of freshwater acidification caused by anthropogenic sulphur and nitrogen emissions, it has become apparent in the 1990s that lakes and rivers react upon a plethora of impacts of both natural and human origin. Climate change was recognised as the major driving force affecting biogeochemistry (i.e. the chemical and biological cycling of matter) and biodiversity (i.e. species composition and interaction) of whole ecosystems. Lakes that have a more or less distinct boundary and shorter response times than most terrestrial ecosystems were chosen as a model ecosystem to study past environmental changes and to predict future developments.

In this sense, lakes were regarded primarily as mirrors of processes occurring in the atmosphere and the terrestrial catchment areas. There are, however, at least two problems with such a simple approach: firstly, limnology (i.e. physics, chemistry and biology of lakes) does not only and entirely depend on atmospheric and terrestrial processes but on the structures and dynamics of the lake ecosystem itself. Secondly, direct human influences (human settlements, transport, agriculture, fisheries, tourism etc.), especially in large and lowland lakes, may exceed the sometimes faint effects of global changes, comprising temperature trends, long-range transboundary air pollution, precipitation patterns, UV radiation etc. For these reasons, it is necessary to study lakes where one can exclude direct human impacts in the catchment area and relate all observed alterations to global changes or natural fluctuations.

Alpine or high mountain lakes offer several advantages for such studies. First, alpine environments can be found all over the world: they are situated above the timber-line which means ca. 2 000–3 000 m altitude in temperate areas, ca. 4 000–5 000 m in tropical, and 0–1 000 m in boreal or polar regions. Alpine lakes may thus serve as a global observation system. Due to the extreme conditions found in alpine lakes, such as low temperatures, strong radiation, low nutrient concentrations, relatively few species and correspondingly 'simple' food webs, they respond more rapidly and more sensitively to external forces. A third reason for selecting alpine lakes as indicators of global change is their role as biodiversity refugia, based above all on the absence of direct human impacts in the catchment area over longer historical periods. Recently, a new aspect has gained importance, and that is the role of mountain areas for humanity: one-seventh of the Earth consists of mountains where one-tenth of the world population lives. But mountains provide essential resources to more than just 10% of humankind, and among those resources, water is a key factor for human survival and development.

In the following sections I compare the current state of knowledge regarding all major forces influencing alpine environments and especially lakes. Finally I try to come to a synthesis of those interacting forces and processes.

2 ACID RAIN

In the late 1970s it became apparent, firstly in Scandinavia but later on in all industrialised countries, that acid rain was the principal cause of profound changes in lakes and rivers. It took, however, many years to understand effects of and reasons for freshwater acidification and to react properly – in contrast with the problem of eutrophication (i.e. fertilisation of lakes) where only 10 years passed between scientific understanding and practical remedy (Table 2.1). Although both problems have some things in common, acidification and eutrophication differ in many regards: while eutrophication is primarily a problem of individual catchments and lakes with little contribution from long-range transport, acidification is a global phenomenon (see Psenner 2001). While eutrophied lakes have been successfully restored by treatments within the lake and/or the catchment area, liming of individual lakes and catchment areas is only a short-term remedy against acidification. Significant improvements can only be expected if sulphur and nitrogen emissions are reduced on a continental or global scale.

From industrial emissions of acid precursors (initially sulphur, later on nitrogen) and the first occurrence of acid lakes in the UK at the turn of

Table 2.1: Comparison of eutrophication and acidification

	Eutrophication	Acidification
Agent		
Primary	Phosphorus	Sulphuric acid H_2SO_4
Secondary	Ammonium, nitrate	Nitric acid, ammonia, organic acids
Sources		
Natural	Mineral dissolution, dust deposition, decomposition of organic matter, release from lake sediments	Volcanism, plant growth, oxidation of pyrite and ammonium, lightning DMS emissions from oceans
Anthropogenic	Detergents, waste water, tourism, agriculture	Industry, traffic, heating, afforestation
Main transport routes		
	Catchment	Long-range transport
Sites	Lowland lakes	Remote, alpine lakes
Geology	Carbonate, silicate	Silicate
Beginning of impact	1950–1960	1850–1950
Recognition of problems		
	1960	1970
Effects	Anoxia, low water transparency, hydrogen sulphide, reduced iron;	Low pH values, high levels of aluminum and heavy metals, oligotrophication;
	changes in species composition, extinction of sensitive species, high biomass of algae and bacteria	changes in biodiversity, extinction of sensitive species, such as algae, molluscs, crustaceans, insects and fishes
Restoration	Since 1960	Since 1970
Lake	Oxygenation, deep-water renewal, iron or aluminium precipitation, biomanipulation	Liming, eutrophication (addition of organic matter and nutrients)
Catchment	Reduction of loading, land use changes	Liming, land use changes
Global	Reduction of P and N emissions	Reduction of S and N fluxes in the biosphere

the 20th century to the massive death of fish populations since the 1920s and their recent virtual extinction in southern Scandinavia, the history of acid lakes and rivers has come a long way. Scandinavian scientists in the 1970s finally managed to prove that acid rain was the main culprit for chemical changes in lakes and streams which led to the extinction of fish and many other species. The final report of the OECD in 1977 clearly showed the impact of sulphur emission, transport and deposition on freshwater acidification, and was followed by the 'Convention on Long-range Transboundary Air Pollution' in 1983. Although some industrial groups tried to sabotage a concerted action to reduce sulphur emissions – a response which seems to have found parallels in the ongoing discussion about greenhouse gases – the convention was ratified by 20 states in 1985. A follow-up convention on the reduction of nitric oxide emissions was signed by 25 states in 1988/89. While NO_x emissions are still high, sulphur emissions have dropped by more than one-third in the industrialised regions during the last 20 years. For many parts of the USA, however, where we saw a 40% reduction in SO_x emissions since 1973, the actual sulphur emissions are still 80% too high (Krajick 2001). This means that sensitive ecosystems will not recover from acidification. Europe emitted 38 million tons of sulphur in 1990 and is expected to reduce that figure to 14 million tons by the year 2010. In the same period, however, sulphur emissions in Asian countries are expected to rise from 34 to 78 million tons per year. Acid rain may thus become less of a problem for North America and Europe but it will certainly remain a major concern for huge areas in Asia throughout the coming decades.

In the early stages of acidification research, a rather parochial approach was quite common, attributing all changes in aquatic ecosystems to specific human activities, for instance SO_x emissions, which were – and in some areas still are – a major threat to hundred thousands of sensitive lakes and rivers all over the northern hemisphere. Consequently, acidification of freshwater was *the* concern for limnologists during most of the 1980s, and the study of acid lakes was pivotal for our understanding of global biogeochemical cycles.

What does acidity mean? A water is called acidic when it contains measurable amounts of protons or H^+, i.e. when the pH drops under a certain value. Another definition points at the balance between base cations (BC = calcium, magnesium, sodium and potassium) and acid anions (AA = sulphate, chloride and nitrate): if the balance is positive, the water contains alkalinity (ALK). This means it is able to buffer the addition of acids.

The difference between BC and AA corresponds more or less to the amount of the hydrogen carbonate or bicarbonate ion (HCO_3^-), thus

$$ALK = BC - AA = HCO_3^- \qquad (2.1)$$

If the input of sulphuric or nitric acid (H_2SO_4 or HNO_3) exceeds the amount of bicarbonate (i.e. the balance BC–AA becomes zero or negative), all HCO_3^- is transformed into CO_2 that will escape to the atmosphere. Consequently, the carbonate-CO_2 buffer system is exhausted, and the water will become acidic, i.e. the pH sinks. A pH of 6 means that the concentration of H^+ is 10 times higher than at neutral conditions (pH 7), pH 5 means 100 times more H+ and so on. When the water is in equilibrium with the CO_2 of the atmosphere, the point of transition between alkalinity and acidity is at pH 5.65. An increase in the CO_2 content of the atmosphere and – consequently – of the concentration of carbonic acid in water will bring down the zero alkalinity pH level. This threshold was considerably higher in preindustrial times when the CO_2 content of the atmosphere was 280 ppm – compared to the present-day value of 360 ppm. In a world without CO_2, the transition pH would be 7, i.e. neutrality.

Since H^+ is a major ion for almost all chemical reactions and many organisms, lowering of the pH can have tremendous consequences on biogeochemical cycles and species composition. One important reaction in acidic water that largely influences its toxicity is the solubility of aluminum hydroxide; $Al(OH)_3$, an end product of the weathering of feldspars, is insoluble at neutral pH but dissolves rapidly if the pH drops below 5.5. It represents, thus, a buffering system acting at low pH values (when the carbonate-CO_2 buffer system is exhausted) that prevents H^+ concentrations from becoming too high. The reaction products, however, consisting of dissolved aluminum ions such as Al^{3+}, $Al(OH)^{2+}$ and $Al(OH)_2^+$, are toxic for plant roots, animals and humans (see Psenner and Catalan 1994).

Alkalinity, pH and the occurrence of aluminum ions provide a clear definition of acidification in a chemical sense: natural waters become acidic when the input of acids from the atmosphere (i.e. from natural and anthropogenic oxidation processes and volcanism) exceeds the input of bases resulting from the dissolution (weathering) of minerals. Globally speaking, we still have an excess of weathering products (base cations) compared to acid anions, and as a consequence 99% of all waters, i.e. the oceans, are alkaline. Nonetheless, acid rain is a global problem in regions with low weathering rates, i.e. in crystalline areas where the bedrock consists of siliceous minerals such as quartz, gneiss and granite (Mosello et al. 1993). Any process – be it natural or anthropogenic – that shifts this balance, influences the state of waters. Afforestation, for instance, or bad management of forests, leads to an unbalanced uptake of base cations by plant roots, consequently the difference between BC and AA may become negative. In contrast, anaerobic conditions lead to a microbial reduction of the acid anions sulphate and nitrate, thus shifting the acid-base balance towards alkalinity.

While the transition from alkaline to acidic conditions is clearly defined at pH 5.65, this is not so for most organisms: molluscs and crustaceans are among the most sensitive species, some of them cannot tolerate pH values below 6.0–6.5, and the most conspicuous changes in species composition occur at pH values between 6.5 and 5.5. So a lake with a pH of 5.8 – still containing alkalinity by chemical standards – may be regarded as severely acidified if one looks at organismic changes. Although there are several fish species which can stand pH values as low as 4.5 as adults, reproductive stages are usually more sensitive: so the presence of many large fish – sometimes regarded as a sign of a good population structure – can be totally misleading and is, in many acidic lakes, the result of reproductive failure. Within one single group of organisms, tremendous differences in sensitivity to low pH can exist. Planktonic diatoms, for instance, are generally extremely sensitive to acidification and disappear at pH values around 6.0–6.5 while benthic forms (i.e. those living on stones or the bottom of lakes) can also tolerate very acidic conditions ('acidophiles'), although some others can live only under alkaline circumstances ('alkaliphiles'). As a result of numerous acidification studies, our knowledge about benthic diatoms and their ecological preferences (pH, alkalinity, dissolved organic carbon, salinity ...) has greatly improved. Actually we possess an excellent indicator system based on the preference of single diatom species. Since their siliceous shells are preserved for centuries in lake sediments, past environmental conditions can be inferred with astonishing accuracy: the diatom inferred pH, for instance, can be measured nearly as precisely as with pH meters. Analysis of benthic diatoms and dating of sediment layers by natural isotopes, such as ^{210}Pb and ^{14}C, allow us to reconstruct lake water pH values for several centuries.

Hundreds of pH reconstructions of sensitive lakes yielded similar trends, i.e. a decrease in pH around 1900 or even before, for instance in England, Wales and Scotland (Battarbee 1986), or between 1930 and 1970, for instance in Scandinavia or in the Alps. With the exception of

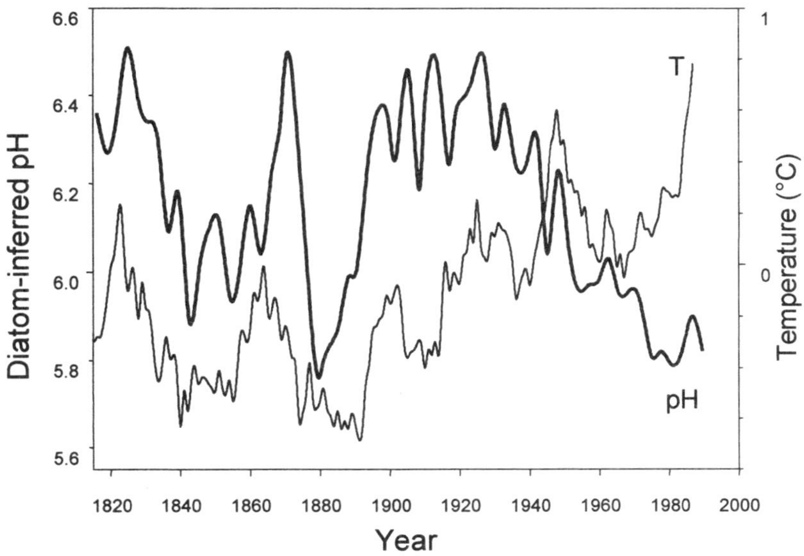

Source: Modified from Psenner and Schmidt (1992).

Figure 2.1: Reconstructed pH of Rasass-See, Vinschgau, South Tyrol and instrumental temperature curves for Alpine stations

Renberg (1990) who reconstructed the pH history of Swedish lakes throughout the Holocene (the last 12 000 years), most authors assumed that acid rain was the only (or at least the major) driving force for pH changes in sensitive lakes. It was, therefore, astonishing to see pH oscillations in two high mountain lakes of the Alps which were of similar amplitude before the onset of acid rain and obviously correlated to changes in air temperature (Figure 2.1).

Koinig et al. (1998) who found comparable oscillations in alpine lakes differing in alkalinity, point out that temperature peaks are well seen even after the onset of acid rain. In the case of Schwarzsee ob Soelden, climate warming was the dominant driving force, exceeding by far the effects of acid rain (Figure 2.2). A comparison of diatom-inferred pH values with electrochemical pH measurements during the last 16 years showed not only a good correspondence of both methods but also the tremendous increase since 1985 which reflects the strong trends in climate warming in the Alps since that time (ca. 1°C, Beniston et al. 1997; see Figure 2.8)

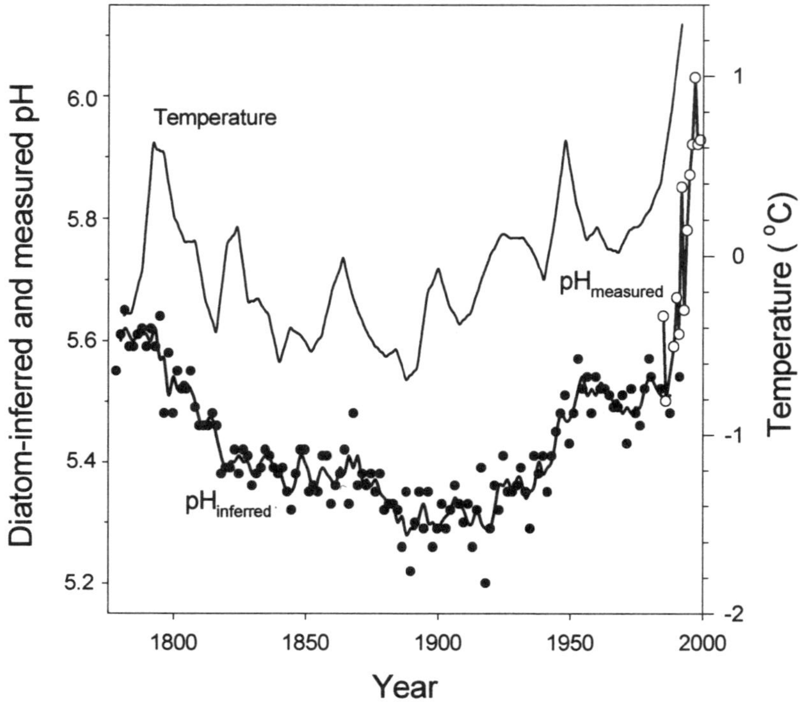

Figure 2.2: Diatom-inferred pH (filled circles) with running average (n = 3, thick line) from 1780 to 1991, and measured pH (open circles) with running average (n = 3, thin line) from 1985 to 2000 in Schwarzsee ob Soelden, Tyrol. Temperature trend at alpine stations (15-year average, thin line)

3 METALS, NUTRIENTS, FLY ASH PARTICLES AND ORGANIC POLLUTANTS

The concentration and the toxicity of metals in lakes is strictly linked to their alkalinity or acidity. At low pH, the solubility and availability of several metals (not only aluminum, see above) increases. Because the deposition of heavy metals has increased with air pollution and acid rain, metal concentrations can reach toxic levels in acid lakes. It is not only the absolute value but the ratio between heavy metals and base cations, especially calcium, that determines their uptake in fish (Figure 2.3).

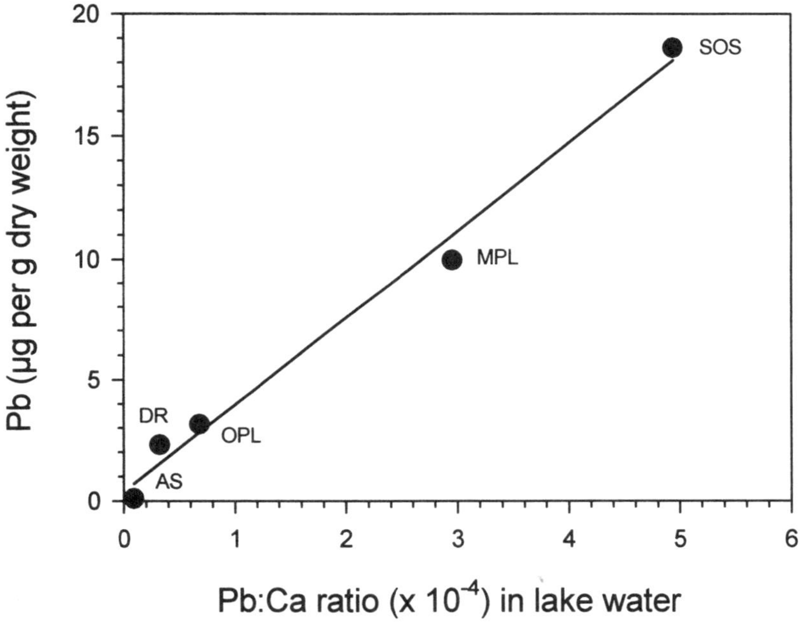

Notes: SOS = Schwarzsee ob Soelden, MPL = Mittlerer Plenderlesee, OPL = Oberer Plenderlesee, DR = Drachensee, AS = Achensee.

Source: Modified from Hofer (1998).

Figure 2.3: Accumulation of lead in the kidney of arctic char from mountain lakes of Tyrol depending on the lead to calcium ratio of lake water

Lead is a special case because its main source was organic lead in petrol, which peaked in the 1980s. Since that time, such additives have been banned and the reduction in deposition can be seen – although blurred by the averaging of several sediment layers – in the sediment core of Gossenkoellesee (Figure 2.4). While another heavy metal, chromium, has been increasing, spheroidal carbonaceous particles (SCPs, or soot particles, generated by high temperature burning of coal and oil) show a threshold around 1850, a steady increase around 1900 until 1930 and a distinct peak around 1960. Since that time, soot particles have dropped to numbers found at the beginning of the 20th century but are still far from reaching background levels.

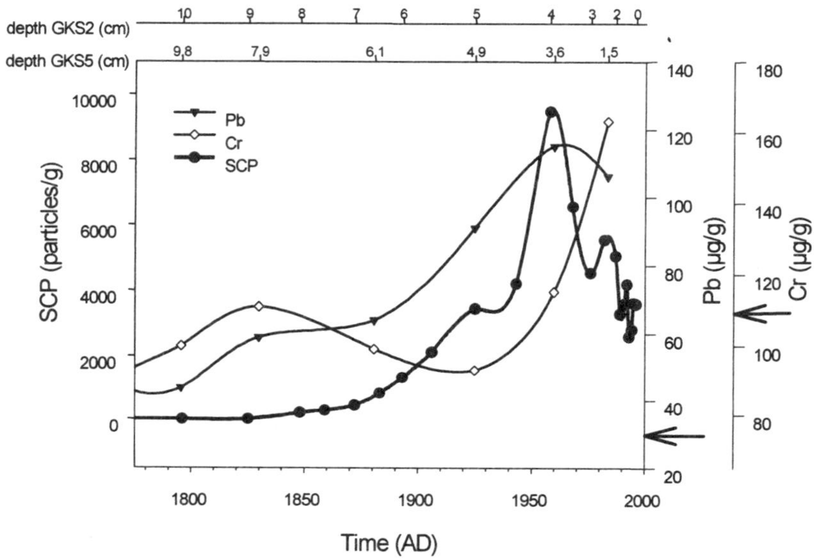

Notes: The cores were dated by use of ^{210}Pb. SCP = soot carbonaceous particles (main source: fossil fuel combustion); Pb (main source leaded gasoline); Cr (sources: coal burning, metal industry); Arrows: background or bedrock concentration of lead and chromium.

Source: Modified from Koinig et al. (2001).

Figure 2.4: Sediment record of atmospheric pollution by soot particles, lead and chromium at Gossenkoellesee, Tyrol

The case of nutrients is more intriguing because at present we lack reliable measurements on phosphorus depositions, which is the limiting nutrient in almost all lakes. Nitrogen, which occurs mainly as nitrate and ammonium in rain and snow, is both an acid and a nutrient, mostly for terrestrial vegetation however, not for lakes. While ammonium (NH_4^+) acts as a base in the chemical balance of waters by buffering, to a large extent, the acidity of rain, it behaves like an acid when it comes into contact with soils, lakes and rivers: ammonium is rapidly taken up by microorganisms and plants (and partly oxidised to nitrate). With respect to our acid-base balance (1) it is obvious that this process creates acidity by removing a base cation, i.e. ammonium (and by producing an acid anion, i.e. nitrate). We can, therefore, say that acid rain looks more neutral than it acts in a terrestrial or aquatic environment.

Regarding the specific situation of alpine catchments, inorganic acids, nutrients and metals are significantly less concentrated in high altitude snow and rain. This is a consequence of both the altitude-dependent increase in precipitation (i.e. a dilution effect) and the dimension of the tropospheric boundary layer which efficiently inhibits the transport of air pollutants beyond ca. 2 000 m altitude, however, only during the winter season. Therefore in terms of inorganic substances, the winter snow in the Alps is almost as clean as in Greenland (Wagenbach pers. comm.). The situation changes completely when we look at the behaviour of persistent organic pollutants or POPs, a class of anthropogenic contaminants with a generally high degree of toxicity and a long half-life in the environment (months to years). Certain POPs, such as DDTs, polychlorinated biphenyls (PCBs), chlordanes, hexachlorocyclohexanes (HCHs), hexachlorobenzene (HCB) and toxaphene are extremely recalcitrant, bioaccumulative and dispersed worldwide. In contrast to inorganic pollutants, they are lipophilic, volatile at ambient temperatures but condensate at low temperatures. They undergo a global distillation and cold trapping process and accumulate, after many evaporation and condensation steps, at the coldest sites of the globe, in polar regions. While contamination of food webs and especially of top predators in the Arctic and Antarctic is known for DDT, DDE and some PCBs, the tremendous increase of those compounds in fish from high altitude lakes in Europe is a new finding (Grimalt et al. 2001).

Transport and accumulation of organochlorine compounds has been investigated in fish and sediments from lakes situated between 430 and 2 800 m above sea level, distributed between 40 and 67°N in Europe. Altitude was the major variable which significantly correlated with fish concentrations and sediment inventory of low volatility species such as 4,4'-DDE, 4,4'-DDT, penta- to heptachlorobiphenyl congeners, i.e. compounds with sub-cooled liquid vapour pressure <0.032 Pa (Figure 2.5). These compounds were significantly correlated with annual average air temperatures at the study sites, thus global distillation of organochlorine compounds in temperate areas results in a selective retention of low volatility substances at the coldest sites. This fractionation effect is responsible for the accumulation of high amounts of potentially harmful compounds in high altitude ecosystems. Arctic char from a lake situated at ca. 2 800 m above sea level had ca. 1 000 times more DDT than those from a lake of ca. 400 m altitude, differing by ca. 8°C in annual mean air temperatures.

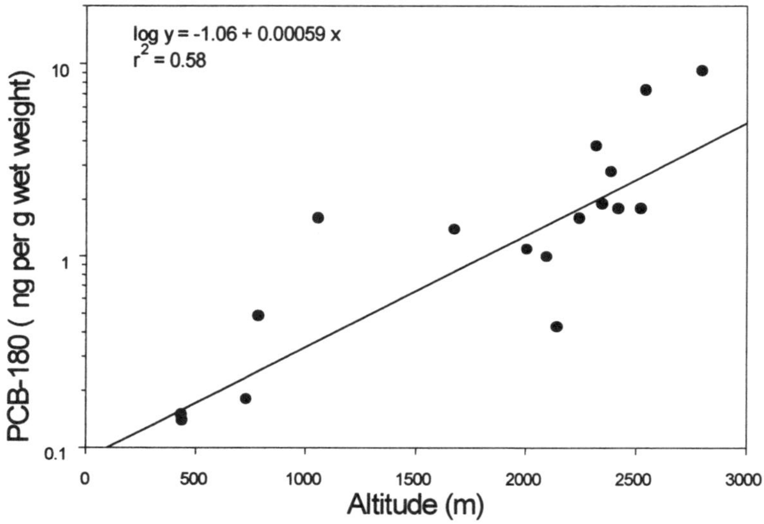

Source: Modified from Grimalt et al. (2001).

Figure 2.5: Concentration of the polychlorinated biphenyl compound PCB-180 (ng pro g wet weight) in salmonid fishes from mountain lakes in Europe depending on the altitude of the lake

4 SOLAR ULTRAVIOLET RADIATION

Solar ultraviolet-B radiation (UVB, 290–320 nm) has increased during the last 15 years over many locations of the Earth as a consequence of the degradation of the stratospheric ozone layer. Besides Antarctica, where the increment is notorious, some studies have demonstrated that a moderate increase in UVB also occurred in the northern hemisphere, for example in the Alps (Blumthaler and Ambach 1990). UVB differs from UVA radiation (320–400 nm) because only the short wavelengths of UVB are energetic enough to damage the DNA molecule while UVA can have secondary effects, for instance on membranes. Moreover, only UVB has increased because of ozone destruction, but still much more UVA than UVB radiation reaches the Earth's surface. Because UVB increases by ca. 10–20% (depending on the wavelength and the location) with each kilometre altitude, UVB radiation at 3 000 m can be up to 50% higher than at sea level.

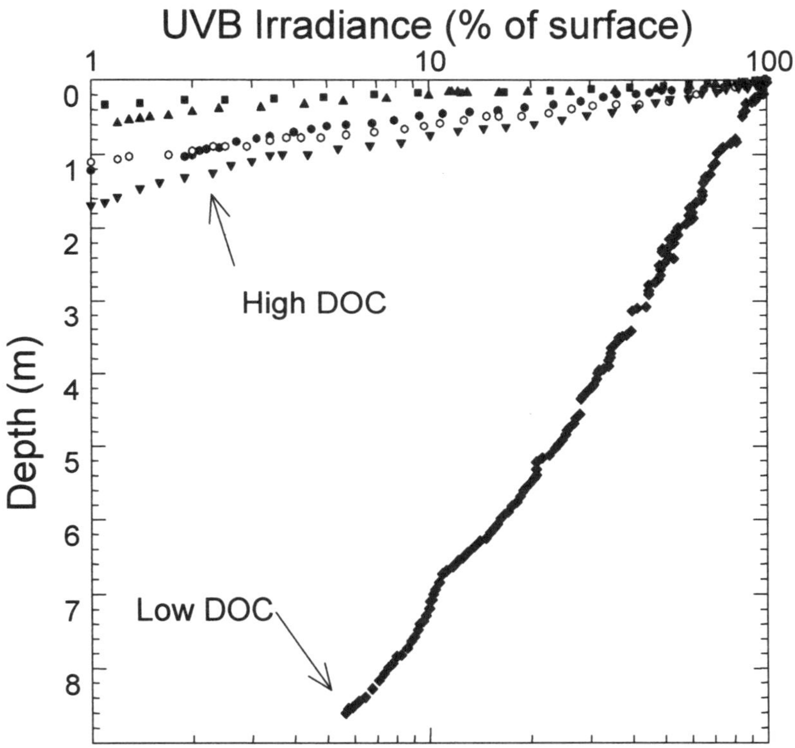

Notes: The depth at which 99% of the incoming radiation is absorbed (1% depth of surface radiation), is between 0.3 and 1.7 m in lakes surrounded by forests, but ca. 16 m in an alpine lake situated above the tree line (Gossenkoellesee, 2 417 m a.s.l.). Since this lake is 9.9 m deep, the whole water column and the bottom are exposed to strong UV radiation.

Source: Modified from Sommaruga and Psenner (1997).

Figure 2.6: Penetration depth of UVB radiation measured at a wavelength of 305 nm in lakes with high and low contents of dissolved organic matter, analysed as dissolved organic carbon (DOC)

The biological response of aquatic organisms to UV radiation is highly variable and non-linear across the UV waveband (i.e. the smaller the wavelength the higher the effect), therefore detailed information on UV exposure as a function of wavelength and water depth is needed to predict the effect of UV radiation in natural environments. Since pure water does not absorb much UV radiation, the chemical and biological composition of lakes and streams is crucial for their capacity to attenuate UV. The most important protective 'substance' is dissolved organic matter (DOM) and, more specifically, chromophoric dissolved organic matter (CDOM or Gelbstoff) deriving from decomposing plant material ('humus', Figure 2.6).

Also chlorophyll *a* (an indicator of phytoplankton biomass) and suspended particles, deriving for instance from glacial flour ('Gletschermilch'), can absorb UV radiation. Dissolved organic matter is the main factor responsible for the variation in UV attenuation among lakes, and DOC (dissolved organic carbon) was found to be a good estimator of UV transparency. The higher the concentration of dissolved organic matter, which may reach up to 30 mg of DOC per litre in humic or brownwater lakes, the shorter the penetration depth of UV radiation and the higher the protection of sensitive organisms such as tadpoles, toads and newts which lack melanin pigmentation (Nagl and Hofer 1997; Pounds 2001). However, dissolved organic matter is significantly reduced in the open ocean and in clear water systems such as polar (Vincent et al. 1998) and alpine lakes (Sommaruga and Psenner 1997). Figure 2.6 shows a typical profile from an alpine lake compared to other lakes with higher contents of DOC: it is evident that alpine lakes are more or less transparent to UV radiation which may therefore even reach the bottom of the lake.

In transparent marine and freshwater systems, phytoplankton ('algae', often measured as chlorophyll *a* can attenuate UV radiation. Sommaruga and Psenner (1997) suggested that the UV attenuation in a high mountain lake during the formation of the deep chlorophyll a maximum was caused also by phytoplankton or substances released by phytoplankton – which function as UV absorbers especially in lakes with low DOC content. To protect themselves from radiation damage, one strategy of phytoplankton is to increase the concentration of photoprotective compounds or 'sunscreens'. An important group of compounds, the mycosporine-like amino acids (MAA), has been found in a wide variety of marine organisms (Karentz et al. 1991), presumably synthesised by algae and accumulated in animals from their food source. These compounds have recently been detected by Sommaruga and Garcia-Pichel (1999) in phytoplankton and zooplankton of high mountain lakes. MAAs are water-soluble, low molecular weight compounds having absorption maxima ranging from 310 to 360 nm and may be released by phytoplankton (Figure 2.7).

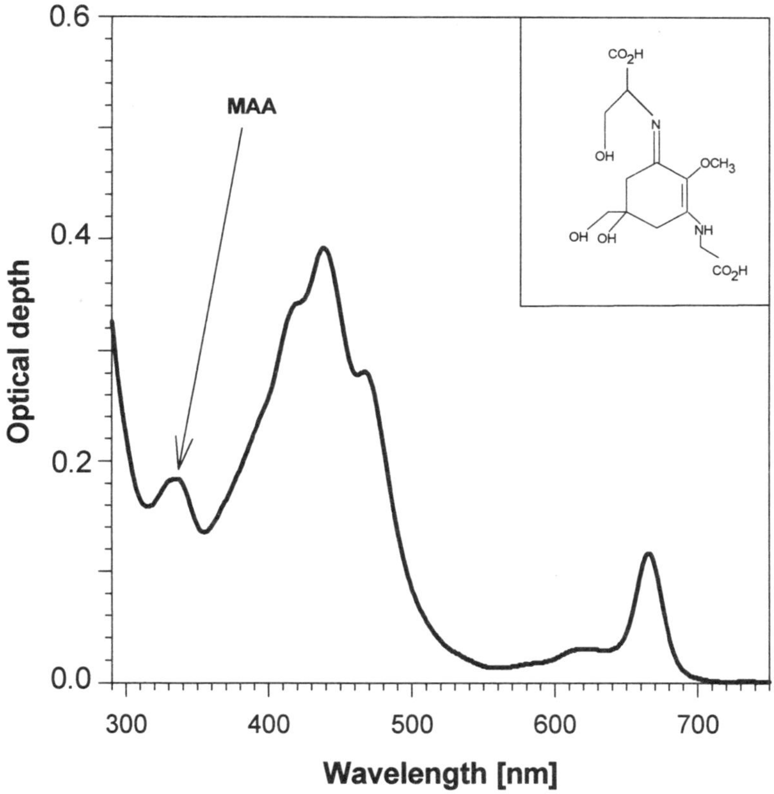

Notes: Inset: structural formula of the MAA shinorine. Absorption peaks at 440 nm and 670 nm are caused by algal photosynthetic and photoprotective pigments, the increase in optical depth at wavelengths below 300 nm is caused by the solvent (90% methanol).

Source: Modified from Sommaruga and Garcia-Pichel (1999).

Figure 2.7: Absorption maximum at ca. 330 nm wavelength of a methanol extract of plankton from Gossenkoellesee (2 417 m a.s.l.), caused by mycosporine-like amino acids (MAA)

The interaction of UV radiation with oxygen and various organic compounds can result in the production of toxic intermediates that can be more damaging than the UV exposure itself. All algal cells contain carotenoids, which are photosynthetic accessory pigments that play a variety of physiological roles. Certain carotenoids, which can also be found in many zooplankton species, can react with and thereby neutralise singlet-state oxygen, confer direct protection of photosystem II reaction centres against photooxydation, and act as general radical-trapping antioxidants. The zooplankton in clearwater lakes evolved different strategies to avoid radiation (e.g. by vertical migration into deeper water layers during daylight), to repair DNA damage (e.g. by photoreactivation) or to protect themselves from high UVB radiation (e.g. by increasing MAA and melanin). Some MAA may function as a biological antioxidant as well. MAAs have been detected in the marine copepod Calanus propinquus (Karentz et al. 1991), the freshwater copepod Cyclops abyssorum tatricus (Sommaruga and Garcia-Pichel 1999) and in other copepods of high mountain lakes (Cabrera et al. 1997; Tartarotti et al. 1999).

5 GLOBAL WARMING

The warming of the Earth's surface over the past 100 years (and a concomitant cooling of the stratosphere) has raised enormous attention, because its effects will change the conditions of most ecosystems and, consequently, everybody's life. I will not argue about the anthropogenic versus natural causes of climate warming, but I want to point to a fact which is of major importance for the further development of lakes in the Alps: from 1895 (the coldest year in the past two centuries) to 1985, the mean air temperature at mountain stations in the Alps has increased by ca. 1°C (Figure 2.8).

Since then, however, we have seen a tremendous increase of ca. 1°C in the last 15 years. Warming in the Alpine area is thus much more pronounced than the global average, and it may have more severe consequences at altitudes with annual average air temperatures around 0°C than at sites with higher annual means because it implies drastic changes in the aggregation state of water: melting of permafrost soils, glacial retreat, disappearance of permanent snow fields, duration and thickness of ice covers, stratification and mixing of the water body, flushing rates etc. will all be influenced massively by a shift of 1–2°C. In addition, weathering rates (depending – among other things – on the presence of liquid water), plant growth and soil development are determined by air temperatures. An increase by 2°C means an upward shift of the snow line (and the boundary between the alpine and the nival zone) by several hundred metres. We will thus see new alpine lakes at higher altitudes, originating at the end of retreating glaciers and snow fields. At the same time, however, some of the alpine lakes situated at around 2 000 m altitude will have a forested catchment area, definitely changing their character over the long run (see section 4 on DOM and UV radiation). Generally, warming – unless combined with drier weather – will enhance productivity, organic matter content, and alkalinity of lakes by a series of factors in the catchment area and the water body itself (Figure 2.9). Warming may, therefore, be regarded as the big competitor with acid rain. The latter pushes lakes in the opposite direction by starting an acidification–oligotrophication process which decreases the concentration of dissolved organic matter, thus having major implications on underwater UV radiation.

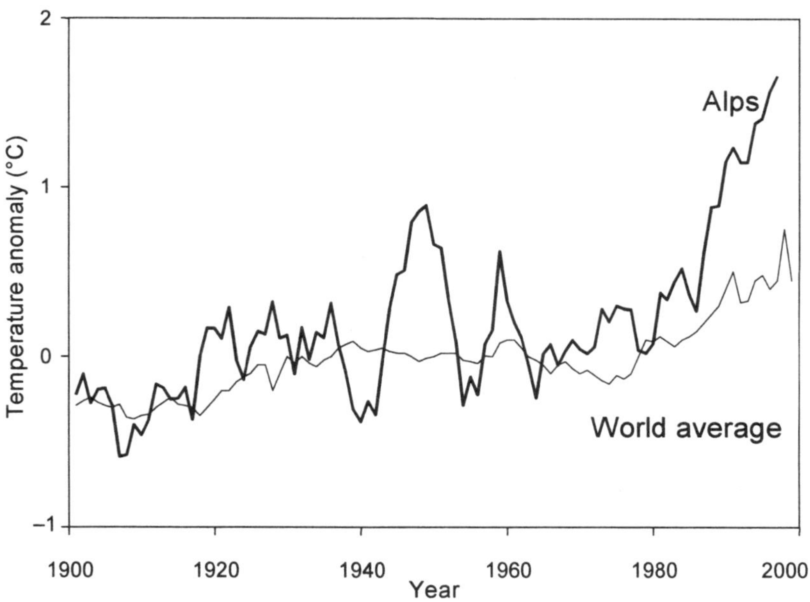

Source: Modified from Beniston et al. (1997).

Figure 2.8: Mean air temperatures; deviation from 30-year mean values for world average (thin line) and Alpine stations (thick line)

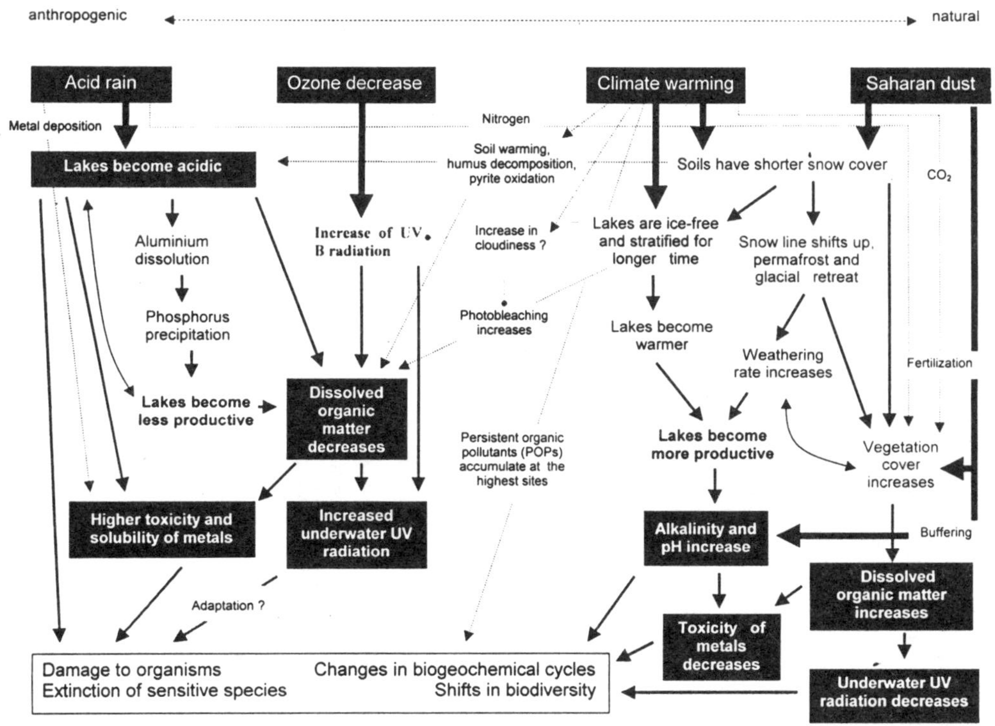

Figure 2.9: Global change effects on alpine lakes

Notes:
Complex interactions exist between acidification, increase in UVB radiation, input of toxic substances, climate warming and Saharan dust deposition. Nutrient, metal and POP deposition is not shown explicitly. On the left-hand side the negative aspects are depicted while the right-hand side shows positive aspects of global changes. An increasing natural component is assumed from left to right.

Bent arrows indicate a positive feedback, for instance between oligotrophication (decrease in productivity) and acidification of lakes as well as between vegetation cover and weathering rates. The degree between natural and anthropogenic influences is illustrated by the horizontal arrow on top.

Crossovers (i.e. factors reverting one effect of acid rain or climate warming), depicted mainly in the middle of the graph, include, for instance, the double role of nitrogen (nutrient and acid) or the photobleaching of dissolved organic matter in open lakes.

Acids (sulphuric and nitric acid) originate from coal and oil combustion but also from natural processes. Low pH means direct toxicity but also increased solubility of toxic metals, especially aluminum, which is insoluble at neutral pH. *UVB radiation* increases with altitude and with ozone destruction. It damages organisms and decomposes dissolved organic matter (DOM). DOM, especially of terrestrial origin (humic acids), absorbs UV radiation, thus reducing the penetration depth of UV, but it also detoxifies heavy metals and aluminum by complexation. Climate warming may have additional negative effects when it is coupled with drier conditions which lead to sinking groundwater levels: as a consequence, oxidation of pyrite (FeS, iron sulphide) and organic nitrogen in soils produces acidity. In most alpine catchment areas, however, climate warming increases weathering rates and productivity of lakes, which generally means higher alkalinity and higher pH. An increase in plant cover in the catchment provides more DOM to the lake water and thus a decrease in the penetration depth of UV radiation. Vegetation growth is favoured also by the deposition of nitrogen which is not only an acid but also a nutrient. At the same time, however, climate warming may increase underwater UVB radiation because the ice cover duration becomes shorter. On the other hand, a warmer and wetter climate may increase cloudiness and thus reduce the incoming radiation. Dust from dry areas of the world (Sahara), but also from industry, road traffic and agriculture, buffers acids and provides nutrients. This adds to the increase in productivity by facilitating warmer and more stable (stratified) conditions of the water column, which favours oxygen depletion at the sediment-water interface and, consequently, the release of phosphorus and the reduction of acid anions (sulphate, nitrate). In addition to the influences outlined in this schematic, one has to consider that 'acid rain' is generally also 'nutrient rain' because it is rich in *nitrogen* compounds (nitrate, ammonia and organic nitrogen). It may have, under certain circumstances, similar effects as dust depositions, i.e. fertilisation, thus favouring plant growth. Also the deposition of *metals and persistent organic pollutants* has increased. The behaviour of persistent organic pollutants differs from that of inorganic pollutants (acids, metals), which generally decrease in concentration with altitude, while large organic molecules such as certain polychlorinated biphenyls (PCBs) and DDT undergo a global distillation process and are trapped at cold sites, for instance in polar regions and high altitudes, where they accumulate. So while high altitude generally means less pollution by acids and metals, it implies increased pollution by certain POPs. In addition to emission levels, accumulation and dispersal of POPs depends primarily on temperature changes.

6 SAHARAN DUST

Although the Sahara has been an enormous source of minerals for ca. 12 000 years, its relevance to alpine lakes has been recognised only recently (Psenner, 1999). Annual dust production in the Sahara is between 60 and 250 Tg (1 Tg = 1 million tons), assumed to be the equivalent to one-quarter of global dust production. For comparison, Sokolik and Toon (1996) estimate a total dust production rate of about 3 000 Tg yr^{-1} with an uncertainty range of 1 000–5 000 Tg yr^{-1} and an anthropogenic contribution of approximately 20% (10–50%). Ca. 10 Tg of dust are exported from the Sahara to Europe per year with a deposition maximum at 2 000–3 000 m altitude. In France, values of 0.5 g m^{-2} per event have been described as common for muddy rains, with maxima of up to 6 g m^{-2}. De Angelis and Guidichet (1991) who studied the dust deposition at Mont Blanc, France and Italy, for a period of 30 years found the Saharan dust to be dominant compared to other continental aerosols: calcium and aluminium inputs from local sources amounted to only 30% and 10% of total input, respectively. They also stated that the high calcium content of the Saharan aerosols might be of importance for biogeochemical cycles involving acid neutralisation. An ice core from Colle Gnifetti, dating back to 1600 AD, revealed the enormous contribution (70%) of Saharan dust to total calcium inventory (Wagenbach et al. 1996). The extremely high $CaCO_3$ content of those dust layers decreased the present acid load by ca. 60%. At a high altitude site in the Central Alps (Hintereisferner, Tyrol, Austria), one Saharan dust event was able to neutralise several metres of acidic snow pack corresponding to accumulation periods of six months or more (Psenner and Nickus 1986). Loye-Pilot et al. (1986)

demonstrated the enormous impact of Saharan dust on rain acidity in the Mediterranean, and also Rodá et al. (1993) and Avila (1996) found that dust depositions, mainly of Saharan origin, effectively buffer acidic depositions at Montseny, Spain. Dust is responsible for the most important difference in precipitation chemistry and, consequently, in the acid-base balance of mountain lakes across Europe (Table. 2.2).

Table 2.2: Mean concentrations of major ions (μeq l^{-1}) in precipitation and high mountain lakes in the Alps and Pyrenees

	Cl^-	SO_4^{2-}	NO_3^-	NH_4^+	Na^+	K^+	Ca^{2+}	Mg^{2+}	BC–AA
Precipitation									
Alps	9	38	22	27	11	4	30	6	–45
Pyrenees	19	50	20	22	23	9	89	7	+17
Lake water									
Alps	7	145	16	2	16	11	331	75	263
Pyrenees	12	42	10	2	25	8	168	15	150

Note: The data are based on analyses of 25 precipitation stations and several hundred softwater lakes in the period from 1985 to 1993. BC = base cations, AA = acid anions (including NH_4^+).

Source: Modified from Camarero et al. (1995).

Rain and snow are acidic in the Alps, but neutral or alkaline in the Pyrenees because of the much higher concentration of base cations (Camarero et al. 1995). In contrast to the Alps, no acidic lakes were found in the Pyrenees, although in both areas lake water alkalinity has decreased because of sulphur and nitrogen emissions. Saharan dust, however, does not only buffer acidic depositions, it can also lead to a significant increase in sulphate and nitrate deposition because basic aerosols take up much more SO_2 and sulphuric and nitric acid. Dust particles are also excellent collectors of organic acids, leading to high acetate and formate levels in snow (Maupetit and Delmas 1994). Dust deposition trends in Europe and North America have been showing a decrease for many years, most probably caused by emission controls. Against this, global dust production is expected to increase because of population growth.

7 TOWARDS A SYNTHESIS

It has been mentioned in the sections above that strong links exist between different aspects of global changes (Yan et al. 1996; Sommaruga-Woegrath et al. 1997; Vinebrooke and Leavitt 1998). These interactions are summarised in a conceptual model (Figure 2.9) that shows, firstly, a degree of human influence on major driving forces – from clearly anthropogenic ones, such as acid rain (which has, however, also some natural sources), to definitely more natural ones, such as dust deposition. Secondly, we can anticipate two 'simple' results; a 'good' one on the right-hand side, and a 'bad' one on the left. There are basically two chains of events (with some positive feedback loops) which lead to a runaway effect with only two possible outcomes: the lakes may gain in productivity, alkalinity and UV protection, or they may become acidic, oligotrophic and transparent to UV radiation, depending on the strength of the four driving forces considered. Climate cooling, for instance, in combination with acid rain will definitely drive alpine lakes into acidification–oligotrophication.

There are, however, several crossovers (outlined mainly in the centre of the graph) which make the whole model more ambiguous or dependent on particular assumptions. While Schindler et al. (1990, 1996) have shown that lakes in Canada have reacted with a burst of acidity in warmer and drier weather conditions, caused, among other factors, by pyrite oxidation in soils, a combination of warmer climate with more humid conditions may have an opposite result. In addition, we have to consider the depth and the organic content of the soil. The role of precipitation has not been considered as extensively as temperature in climate scenarios because it is much less predictable and more patchy than temperature distribution. Another question regards the double role of nitrogen compounds which behave as acids and nutrients at one and the same time, and the predicted increase in cloudiness, which may prevent the enhancement of photodegradation of organic matter, as anticipated under the scenario of longer ice-free seasons.

Certainly the most important question is the future of acid precipitation and climate warming, because it is this 'competition' that determines the fate of alpine lakes. If predictions about the doubling of sulphur emissions by growing populations in Asia are correct (see section 2), the frequency and intensity of acid rain will increase over large areas in Southeast Asia. The expected increase in high altitude sulphate aerosols may lead to a cooling effect, and consequently we can anticipate a shift to the left-hand side of the model. If climate warming turns out to be stronger than predicted, it may counteract even increasing acidification of atmospheric depositions. Desert dust may buffer part of this acidity but the degree of neutralisation will depend on future changes in desertification trends and wind patterns. Although we are far from knowledgeable about the details of many interactions, there are some general features which justify the notion that complex interactions may lead to simple effects. So the only suggestion one can give is to reduce the human pressure on ecosystems – and to be prepared for new and unexpected links.

BIBLIOGRAPHY

Avila, A. (1996), Time trends in the precipitation chemistry at a mountain site in north-eastern Spain for the period 1983–1994, *Atmospheric Environment* **30,** 1363–1373.

Battarbee, R.W. (1986), Diatom analysis, in B.E. Berglund (ed), *Handbook of Holocene Palaeoecology and Palaeohydrology*, John Wiley & Sons, Chichester, pp. 527–570.

Beniston, M., H.F. Diaz and R.S. Bradley (1997), Climatic change at high elevation sites: an overview, *Climatic Change,* **36**, 233–251.

Blumthaler, M. and W. Ambach (1990), Indication of increasing solar ultraviolet-B radiation flux in alpine region, *Science,* **248**, 206–208.

Cabrera, S., M. Lopéz and B. Tartarotti (1997), Phytoplankton and zooplankton response to ultraviolet radiation in a high-altitude Andean lake: short- versus long-term effects, *Journal of Plankton Res.,* **19**, 1565–1582.

Camarero, L., J. Catalan, R. Mosello, A. Marchetto and R. Psenner (1995), Comparison of the chemistry of high altitude lakes in central, southwest and southeast Europe (Alps, Pyrenees, Pirin), *Limnologica,* **25**, 141–156.

De Angelis, M. and A. Guidichet (1991), Saharan dust deposition over Mont Blanc (French Alps) during the last 30 years, *Tellus,* **43B**, 61–75.

Delmas, V., H.G. Jones, M. Tranter and R. Delmas (1996), The weathering of aeolian dusts in alpine snow, *Atmospheric Environment*, **30**, 1317–1325.

Grimalt, J.O, P. Fernandez, L. Berdie, R. Vilanova, J. Catalan, R. Psenner, R. Hofer, P.G. Appleby, B.O. Rosseland, L. Lien, J.C. Massabuau and R.W. Battarbee (2001), Selective trapping of organochlorine compounds in mountain lakes of temperate areas, *Environ. Sci. Technol.*, **35**, 2 690–2 697.

Hofer, R. (1998), Fische in alpinen Hochgebirgsseen: Oekotoxikologische und oekophysiologische Aspekte, *Verh. Ges. Ichthyol.*, **1**, 59–73.

Karentz, D., F.S. McEuen, M.C. Land and W.C. Dunlap (1991), Survey of mycosporine-like amino acid compounds in Antarctic marine organisms: potential protection from ultraviolet exposure, *Mar. Biol.*, **108**, 157–166.

Koinig, K.A., R. Schmidt, S. Woegrath, R. Tessadri and R. Psenner (1998), Climate change as primary cause for pH shifts, *Water, Air and Soil Pollution,* **104**, 167–180.

Koinig, K.A., C. Kamenik, R. Schmidt, A. Agusti-Panareda, P. Appleby, A. Lami, M. Prazakova, N. Rose, O.A. Schnell, R. Tessadri, R. Thompson and R. Psenner (2001), Environmental changes in an alpine lake (Gossenkoellesee, Austria) over the last two centuries – the influence of air temperature on biological parameters, *J. Paleolim* (in press).

Krajick, K. (2001), Long-term data show lingering effects from acid rain, *Science*, **292**, 195–196.

Loye-Pilot, M.D., J.M. Martin and J. Morelli (1986), Influence of Saharan dust on the rain acidity and atmospheric input to the Mediterranean, *Nature*, **321**, 427–428.

Maupetit, F. and R. Delmas (1994), Carboxylic acids in high-elevation Alpine glacier snow, *J. Geophys. Res.*, **99**, 16 491–16 500.

Mosello, R., A. Barbieri, G. Bendetta, A. Boggero, A. Marchetto, R. Psenner, D. Tait and G.A. Tartari, G.A. (1993), Quantification of the susceptibility of alpine lakes to acidification, *Mem. Ist. ital. Idrobiol.*, **52,** 355–386.

Nagl. A.M. and R. Hofer (1997), Effects of ultraviolet radiation on early larval stages of the Alpine newt, *Triturus alpestris*, under natural and laboratory conditions, *Oecologia*, **110**, 514–519.

Pounds, J.A. (2001), Climate and amphibian declines, *Nature*, **410**, 639–640.

Psenner, R. (1999), Living in a dusty world: airborne dust as key factor for alpine lakes, *Water, Air, and Soil Pollution,* **112**, 217–227.

Psenner, R. (2001), Gewaesserversauerung, in M. Dokulil, A. Hamm and J.-G. Kohl (eds), *Oekologie und Schutz von Seen*, Facultas Verlag Wien UTB, 251–273.

Psenner, R. and J. Catalan (1994), Chemical composition of lakes in crystalline basins: a combination of atmospheric deposition, geologic background, biological activity and human action, in R. Margalef (ed), *Limnology now: A Paradigm of Planetary Problems*, Elsevier, Amsterdam, pp. 255–314.

Psenner, R. and R. Schmidt (1992), Climate-driven pH control of remote alpine lakes and effects of acid deposition, *Nature*, **356**, 781–783.

Psenner, R. and U. Nickus (1986), Snow chemistry of a glacier in the Central Eastern Alps (Hintereisferner, Tyrol, Austria), *Z. Gletscherkunde u. Glazialgeol.*, **22**, 1–18.

Renberg, I. (1990), A 12 000 year perspective of the acidification of Lilla Oeresjoen, south-west Sweden, *Phil. Trans. Roy. Soc. Lond.*, **B 327**, 357–361.

Rodá, F., J. Bellot, A. Avila, A. Escarre, J. Pihol and J. Terradas (1993), Saharan dust and the atmospheric inputs of elements and alkalinity to Mediterranean ecosystems, *Water, Air and Soil Pollution*, **66**, 277–288.

Schindler, D.W., P.J. Curtis, B.R. Parker, and M.P. Stainton (1996), Consequence of climate warming and lake acidification for UV-B penetration in North American boreal lakes, *Nature*, **379**, 705–708.

Schindler, D.W., K.G. Beaty, E.J. Fee, D.R. Cruikshank, E.R. Debruyn, D.L. Findlay, D.L. Linsey, M.P. Stainton and M.A. Turner (1990), Effects of climate warming on the lakes of the central boreal forest, *Science*, **250**, 967–970.

Sommaruga, R. and F. Garcia-Pichel (1999), UV-absorbing mycosporine-like compounds in planktonic and benthic organisms from a high-mountain lake, *Arch. Hydrobiol.*, **144**, 255–269.

Sommaruga, R. and R. Psenner (1997), Ultraviolet radiation in a high mountain lake of the Austrian Alps: air and underwater measurements, *Photochem. Photobiol.*, **65**, 957–963.

Sommaruga-Woegrath, S., K.A. Koinig, R. Schmidt, R. Sommaruga, R. Tessadri and R. Psenner (1997), Temperature effects on the acidity of remote alpine lakes, *Nature*, **387**, 64–67.

Tartarotti, B., S. Cabrera, R. Psenner and R. Sommaruga (1999), Survivorship of *Cyclops abyssorum tatricus* (Cyclopoida, Copepoda) and *Boeckella gracilipes* (Calanoida, Copepoda) under ambient levels of solar UV radiation in two high mountain lakes, *J. Plankton Res.*, **21**, 549–560.

Vincent, W.F., R. Rae, I. Laurion, C. Howard-Williams and J.C. Priscu (1998), Transparency of Antarctic lakes to solar UV radiation, *Limnol. Oceanogr.*, **43**, 618–624.

Vinebrooke, R.D. and P.R. Leavitt (1998), Direct and interactive effects of allochthonous dissolved organic matter, inorganic nutrients, and ultraviolet radiation on an alpine littoral food web, *Limnol. Oceanogr.*, **43**, 1 065–1 081.

Wagenbach, D., S. Preunkert, J. Schaefer, W. Jung and L. Tomadin (1996), Northward transport of Saharan dust recorded in a deep alpine ice core, in S. Guerzoni and R. Chester (eds), *The Impact of Desert*

Dust across the Mediterranean, Kluver Academic Publishers, Dordrecht Boston London, 291–300.

Yan, N.D., W. Keller, N.M. Scully, D.R.S. Lean and P.J. Dillon (1996), Increased UV-B penetration in lake owing to drought-induced acidification, *Nature*, **381**, 141–143.

3. On the Economics of Climate Change and the Climate Change of Economics

Stefan P. Schleicher

1 INTRODUCTION

Global warming was, for most economists, as observed by Thomas Schelling (1992) in his famous presidential address for the American Economic Association, an unknown territory. Unfortunately, judging by how most economic modeling has been seen by experts outside the field of economics, i.e. as utterly naive, this still appears to be true for the majority of economists today. The striking example is the controversy about the timing of climate policy, where economists tend to recommend waiting, in contrast to climatologists who call for immediate action (Nordhaus 1998).

Economic evaluation of specific events triggered by climate change seems to be even more difficult. When climatologists warned of the loss of coastal land caused by sea level rise some economists blandly answered that these losses will be compensated for by an increase of value in the land behind. Such responses do nothing to raise the stature of the discipline, no matter how correct they may be from the point of view of individual utility maximisation. Climate issues, however, are dealing with intergenerational issues for which the toolbox of mainstream economics may not be well equipped. In this contribution we suggest that the challenge of climate change has become a challenge for the mainstream paradigm of economics. Or, to put it in other words, the phenomenon of ongoing climate change has stimulated research that questions the adequacy of the standard building blocks used in mainstream economics (Nordhaus, 1994; Kratena and Schleicher, 1999; Gupta and Grubb, 2000; Weyant, 2000, Abele et al., 2001). Basically the heated debates about the policy options meeting the challenge of climate change have become a challenge for expanding the horizon of economic reasoning.

As a testing ground for the capability of economic analysis to handle climate issues serves the Alpine region. This environment is likely to be affected by a global climate change not only through the projected increase of average temperatures but also through changes in many other climatic elements. In particular changes in the magnitude and timing of precipitation would have significant impacts not only on the economy and ecology of Alpine regions but would be felt also far beyond the mountainous regions because of changes in the water cycle (Beniston, 1997, 1998, 2000; Buerki, 1998; Cebon et al., 1998).

We therefore start with a brief collection of facts about climate change in Alpine regions, as they are accessible to economists, and then attempt their evaluation using the lines of argumentation of the mainstream paradigm. The obvious deficiencies that result are used to explain why a fundamental switch from the flow-oriented to stock-oriented paradigms seems advisable. Thus, the climate issue might serve as a practical instrument for testing the usability of economic paradigms.

2 SOME FACTS ABOUT CLIMATE CHANGE IN THE ALPS

2.1 Observed and Expected Climate Variations

According to the *IPCC Special Report on the Regional Impacts of Climate Change* (1999) and the *IPCC Third Assessment Report* (2001) the European Alps may be heavily affected by changes in global climate. Up to 95% of Alpine glacier mass could disappear by 2100 with subsequent consequences for the water flow regime. Simulations of climate trends suggest that an increase of temperatures would heavily affect winter tourism even during the next few decades. Precipitation is expected to increase in winter and to decline in summer.

CLEAR (2001), the Swiss National Science Foundation research project on climate change in Alpine regions, has provided preliminary empirical evidence. Since the middle of the last century this shows that temperature records for the Alpine regions reflect a gradual increase of mean annual temperature which is roughly coincident with the observed global warming trend. Superimposed on this long-term increase are stronger but more short-term fluctuations. There are also noticeable changes in precipitation with evidence of increasing trends for the winter season. Major shrinkage has been recorded for the majority of glaciers.

Despite the high degree of uncertainty regarding the physical cause of the observed climatic trends, the variations are consistent with general expectations concerning the impact of an enhanced anthropogenic greenhouse effect (Price, 1999; Visconti, Beniston, Hannorelli and Barta, 2001; Wanner et al., 2000).

These findings suggest that pronounced changes in Alpine climate are occurring: increases of general temperature, weather variability and precipitation during the winter season, coupled with a shift from snowfall to rainfall in areas at intermediate levels.

2.2 Socio-economic Impacts

The most prominent impact of climate change in Alpine regions stems from the relevance of extreme weather events. Avalanches, landslides, flash floods or soil erosion will result from more frequent wind storms, rain or snowfall. Exposed settlements and traffic lines that are vulnerable to these events require relocation or additional investments for protection. Data from the insurance industry (Karl, 1999; see also Chapters 5 and 7 in this book) suggest that damage claims from weather-related catastrophes are increasing in Alpine regions.

More gradual changes from climate change are expected for the mountainous ecosystems. Agricultural activities may suffer from the increased weather variability. Alpine forests may lower their capacity as a natural provider of protection against avalanches and severe precipitation. Of special concern is the impact of climate changes on regional-scale hydrological cycles. The potential threats range from extensive flooding to disturbances in hydropower production.

All these changes require a more coherent and integrated view in order to explain the full impact on economic and living conditions in Alpine regions. Winter tourism, which accounts in many regions for one-third of annual income, may be most heavily hit. Substantial economic losses from snow-deficient winters below areas of 1 200 m will force a shift to higher locations.

3 EVALUATING CLIMATE CHANGE: A TEST OF ECONOMIC PARADIGMS

We take the evidence of an ongoing climate change in Alpine regions as a challenge for testing how economic paradigms cope with such phenomena. In particular, we subject the following expected effects of climate change to an economic evaluation:

- decline in tourism;
- damages from avalanches and flooding;
- safety investments in settlements and infrastructure;
- relocation of settlements;
- retreat of glaciers.

We begin with the paradigms of mainstream economics, which can be characterised as flow-oriented, and then provide arguments for switching to a new class of paradigms that emphasise the role of stocks.

3.1 Stage One: Flow-oriented Economic Paradigms

The starting point for our evaluation of the usefulness of economic paradigms in analysing climate change is the mainstream paradigms: these focus on flows for describing economic activity. In particular, these paradigms emphasise the flows of production q (gross domestic product) and (private and public) consumption c as most relevant for describing the state of an economy.

The basic model

In a nutshell these flow-oriented paradigms can be described by the following three-equation model (3.1). This focuses on the production of products using reproducible resources. By using production technology $q()$, the flow of current inputs r^q and the stock of reproducible resource R^q (equipment and buildings), yield the flow of products q which can be used either for investment into reproducible resources i^q or for consumption c. The stock of reproducible capital is affected by the depletion of current inputs r^q and investments i^q. Welfare W is measured in terms of available flows for consumption c, alternatively by the flow of produced products q.

$$W = W(c) \tag{3.1a}$$
$$c + i^q = q(r^q, R^q) \tag{3.1b}$$
$$R^q = R^q(R^q_{-1}, i^q, r^q) \tag{3.1c}$$

The economy represented by this model is driven by the decision about investments for reproducible capital i^q, with consumption c being the residual left after this decision has been taken, thus putting a limit on actual consumption.

Adding recycling

An important improvement in the framework of this flow-oriented paradigm is the introduction of recycling activities *g()* that enable the reuse of wastes from production or consumption, thus providing the same volume of products for consumption and investments with lower flows of newly produced products *q*. This means that in addition to the stock for reproducible capital R^q, the corresponding investment i^q and input flows r^q, the same interactions have to be taken into account for the capital stock for recycling activities R^g and corresponding input flows r^g and investments i^g;

$$W = W(c) \quad (3.1a)$$
$$c + i^q + i^g = q(r^q, R^q) + g(r^g, R^g) \quad (3.1b')$$
$$R^q = R^q(R^q_{-1}, i^q, r^q) \quad (3.1c)$$
$$R^g = R^g(R^g_{-1}, i^g, r^g) \quad (3.1d)$$

Again the dynamics of the economy represented by this model is driven by the decision about investments for reproducible capital and investments for recycling capital. The relation of these two types of investments determines the extent to which consumption and investments can be supplied from recycling activities.

A first evaluation of climate change

Applying these flow-oriented paradigms for evaluating the economic impacts of climate change in Alpine regions we several revealing results.

A decline in the demand for the consumption of tourism would lower both consumption and production and this decline would be interpreted as a loss of welfare. Damage from avalanches and flooding would surprisingly not show up in the welfare measure of this paradigm although the reproducible stocks R^q are reduced. But the resulting repair activities of rebuilding houses and infrastructure would stimulate investments and raise economic activity.

Depending on restrictions of production capacities and loans, consumption opportunities could indirectly be negatively affected. But in the case of unused production capacities obvious damage can be given even a favorable economic interpretation because of the model's focus on product flows and its neglect of losses in capital stock. The same, rather strange, interpretation would be given to investments in safety, infrastructure and settlement relocation. The impact of glaciers retreat would not be evaluated at all in such paradigms.

3.2 Stage Two: Stock-oriented Economic Paradigms

Economic models or paradigms that put their emphasis on the relevant stocks of an economy lead to a fundamental change in the interpretation of economic events.

Welfare from flows and stocks

As a first step it is observed that welfare results not only from flows but also from (reproducible) stocks. An excellent example is the heating of a room to a comfortable temperature. This needs much lower flows of energy if the thermal structure of the building – the relevant stock – is of higher quality:

$$W = W(c, R^c) \tag{3.2a'}$$

The implication is that the same level of welfare may be maintained with a much lower flow of economic activity, e.g. in terms of production flows. Of course substitution of flows c by stocks R^c depends on the available technologies. Pilot projects for buildings, for example, indicate that energy flows can be almost completely substituted by a capital stock that provides the adequate thermal quality.

Taking emissions into account

In order to obtain a more complete picture of the impact of economic activities we now introduce emissions which result from production (and implicitly also from consumption). We assume that the flow of emissions e increases with the level of production flows q but decreases with the amount of current inputs r^e and capital stock R^e devoted to emission abatement:

$$e = e(q, r^e, R^e) \tag{3.2f}$$

In addition we take into account the stock or concentration of emissions E (e.g. CO_2 in the atmosphere) by observing the absorption capacity of the relevant environment:

$$E = E(E_{-1}, e) \tag{3.2g}$$

Considering natural capital

The final extension of our stock-oriented paradigm involves the inclusion of natural capital N, ranging from the stock of fauna and flora to the stock of glaciers. Two consequences of this perspective have to be observed. First, the stock of environmental resources and the stock of natural capital should be included in our measure of welfare:

$$W = W(c, R^c, E, N) \tag{3.2a}$$

Second, we should attempt to model the interaction of natural capital with the stock of environmental resources (air, water and soil) and economic activity:

$$N = N(N_{-1}, E, q) \tag{3.2i}$$

Thus we end up with the following model that represents the fundamental building blocks of a stock-oriented paradigm of economic activity which is also capable of considering interactions with environmental and natural resources:

$$W = W(c, R^c, E, N) \tag{3.2a}$$
$$c + i^q + i^g + i^c + i^e = q(r^q, R^q) + g(r^g, R^g) \tag{3.2b}$$
$$R^q = R^q(R^q_{-1}, i^q, r^q) \tag{3.2c}$$
$$R^g = R^g(R^g_{-1}, i^g, r^g) \tag{3.2d}$$
$$R^c = R^c(R^c_{-1}, i^c, r^c) \tag{3.2e}$$
$$R^e = R^e(R^e_{-1}, i^e, r^e) \tag{3.2f}$$
$$e = e(q, r^e, R^e) \tag{3.2g}$$
$$E = E(E_{-1}, e) \tag{3.2h}$$
$$N = N(N_{-1}, E, q) \tag{3.2i}$$

This model emphasises the crucial role in the investment decision for four stocks of capital: capital used for new and recycled products, R^q and R^g, respectively, capital that is relevant for the provision of services to consumers, R^c, and capital that is employed for emissions reductions, R^e. These specifications indicate the wide span of economic structures that are compatible with a desired level of welfare, ranging from high-flow to low-flow structures.

The following issues are all addressed in this comprehensive model:

- First, the evaluation of economic welfare is not limited to the flows of consumption (or production) but also considers the stocks of reproducible, environmental and natural capital.
- Second, the flow of production can be lowered by increasing recycling activities.

- Third, flows for consumption and production can be substituted at least to some extent by the corresponding capital stocks (this option was only considered for production in the stage one model).
- Fourth, the fact that environmental and natural capital have only limited absorption capacities with respect to emissions and production activities can be addressed (in terms of the functional form of *N*()).

A refined evaluation of climate change

In the framework of the stage-two model that focuses on stocks of reproducible, environmental and natural capital, we are able to produce a much more constructive evaluation of climate change issues as expected for the Alpine regions.

As in the stage one model declines in tourism show up as direct losses on consumption and damages to producible capital stock as indirect losses of consumption opportunities.

But additional losses also become visible using this more refined measure of welfare. Damages to the capital stock of consumers caused by storms and flooding lower consumer welfare. So does the degradation of environmental resources, above all the increase of greenhouse gases in the atmosphere (e.g. experienced as adverse weather conditions) and the loss of natural resources (e.g. the retreat of glaciers).

Looking at these issues in more detail, we realise that the destruction of reproducible capital stock both with consumers and producers (e.g. by floods and avalanches) has more consequences than just lowering consumption opportunities (the impact already indicated in the stage one model). Replacement of damaged capital stock and additional investments in settlement safety, infrastructure and relocation put limits on the available production capacities for investments in resource efficiency (e.g. recycling) and emissions reductions (e.g. energy efficiency).

In addition it should me mentioned that some highly relevant aspects for evaluating economic activities are still missing in the proposed modeling framework. The interaction of economic activity and cultural identity can be traced in the history of economic thought (Mirowski, 1994) and it seems that the valuable link of economics to the design of culture, at least for the time being, seems to have been lost. As, for example, we are rediscovering the importance of biodiversity of a rain forest versus its economic market value, we may very well come to the conclusion that the cultural identity of people deserves priority in its own right, i.e. beyond any economic reckoning. Another missing item for the evaluation of economic activity is – as again Mirowsky (1989) indicates

– the potential link of economic reasoning to the natural sciences, especially to the importance of the First and Second Law of thermodynamics which provides valuable guidelines for the design of technologies for recycling and cogeneration of heat and electricity.

3.3 Guidelines for Policy Design

This refined modelling framework not only reveals the use of a deficient economic paradigm as the source of many current conflicts (e.g. as seen in the implementation of the Kyoto Protocol), it also serves to identify recommendations for policy design.

The first recommendation relates to the use of GDP as a measure of economic welfare. It is obvious from the refined specification of welfare in (equation 3.2a) that the role of flows should gradually be de-emphasised. In addition, incentives should be provided to aid the substitution of consumption flows by stocks of reproducible capital as long as the efficiency of the whole system suggests to do so.

The second recommendation concerns the role of recycling activities as seen in (equation 3.2b). From waste paper to waste heat there are lots of opportunities to lower material flows without lowering the level of welfare. The level of recycling appears to be too low in present day economies.

The third recommendation deals with the level of emissions which, according to (equation 3.2c), can be extensively reduced by abatement investment. The extent to which this should be done again should be guided by efficiency criteria of the whole system.

The fourth recommendation reveals the crucial role of the choice of technologies for production, recycling and emissions abatement. This choice also involves switching from high to low carbon and ultimately potentially carbon-free, primary inputs, if the challenge of greenhouse impact is acknowledged.

The fifth recommendation concerns changes in the institutional settings that are needed to stimulate structural changes in the design of economic activity along the lines suggested above. Obviously a wide range of options, from stimulating R&D in fossil-free energy technologies to arranging crucial international agreements on emissions, can be employed.

Finally, the insight to preserve the riches of cultural identities provides yet another argument for switching to lifestyles that attempt to maintain cultural foundations of as regions as the Alps.

4 SUSTAINABILITY FOR ALPINE REGIONS

When assessing the foreseeable impacts of climate change on Alpine regions, conventional economic analysis evaluates the influence on tourism and agriculture, the costs of relocating homes and production facilities, and the costs of adaptation. Realising the deficiencies of this approach, we have presented a comprehensive modelling framework that above all expands the concepts of capital and of welfare. We end up with a refined modeling framework that essentially is an attempt to make the concept of sustainable development operational.

The notion of sustainability, first proposed by the so-called Brundtland Commission (WCED, 1987) contains at first glance a very modest normative statement: the proposition to organise economic activity in a way that has no negative impacts on future generations. However, closer inspection soon reveals that the economic processes of production and of consumption, at least in industrialised countries, still fail to meet even such modest proposals.

On attempting to evaluate the economic impact of climate change in the Alps, we discovered the need to monitor indicators for the stocks of environmental resources (as air, soil, water) and for the stocks of natural capital. These indicators are also included in a refined measure of welfare that de-emphasises the relevance of the flow of consumption goods in favour of the services generated by the stock of capital goods.

Finally, it should be mentioned that on applying those indicators to our local economies in Alpine regions, we find that many of the changes which clearly contradict the guidelines of sustainability, are in most cases not the result of ongoing climate changes but of unsustainable practices in the tourist and agricultural industries. Thus, it might turn out that the long-term challenge of climate change might have a very positive impact on the short-term necessities of policy making, in that sustainable economic structures would not only serve the climate in the century ahead but above all be highly beneficial for the way we organise consumption and production today.

BIBLIOGRAPHY

Abele, H., T.C. Heller, S.P. Schleicher, Eds. (2001). *Designing Climate Policy. The Challenge of the Kyoto Protocol.* Vienna: Service Fachverlag.

Beniston, M. (1997). *Climate Change and its Consequences for Mountain Regions*. Vienna: Austrian Council on Climate Change, www.accc.gv.at.

Beniston, M. (1998). *The Impacts of Climate Variability on Forests*. Berlin: Springer Verlag.

Beniston, M. (2000). *Environmental Change in Mountains and Uplands*. Munich: Arnold.

Buerki, R. (1998). Klimaaenderung und Skitourismus - Wie wirkt sich eine Klimaaenderung auf die skitouristische Nachfrage aus? *Geographica Helvetica*, 53(4), 155–161.

Cebon, P., U. Dahinden, H. Davies, D.M. Imboden and C.C. Jaeger, Eds. (1998). *Views from the Alps: Regional Perspectives on Climate Change*. Boston, MA: MIT Press.

CLEAR (2001). Swiss National Foundation Research Project on Climate Change in Alpine Regions. http://clear.eawag.ch.

Gupta, J. and M. Grubb, Eds. (2000). *Climate Change and European Leadership: A Sustainable Role for Europe?* Amsterdam: Kluwer Academic Publishers.

IPCC (1999). *IPCC Special Report on the Regional Impacts of Climate Change*. Geneva: Intergovernmental Panel on Climate Change.

IPCC (2001). *IPCC Third Assessment Report*. Geneva: Intergovernmental Panel on Climate Change.

Karl, T., Ed. (1999). *Weather and Climate Extremes: Changes, Variations, and a Perspective from the Insurance Industry*. Amsterdam: Kluwer Academic Publisher.

Kratena, K. and S.P. Schleicher (1999). Impact of Carbon Dioxide Reductions on the Austrian Economy. *Economic Systems Research*, 11, 245-261.

Mirowski, Philip (1989). *More Heat than Light*. Cambridge, UK: Cambridge University Press.

Mirowski, Philip, Ed. (1994). *Natural Images in Economic Thought*. Cambridge, UK: Cambridge University Press.

Nordhaus, W.D., Ed. (1994). *Managing the Global Commons*. Cambridge, MA: The MIT Press.

Nordhaus, W.D., Ed. (1998). *Economics and Policy Issues in Climate Change*. Washington, DC: Resources for the Future.

Price, M.F. (1999). *Global Change in the Mountains*. New York: Parthenon Publishing Group.

Schelling, T.C. (1992). Some Economics of Global Warming. *American Economic Review* 82(1), 1–14.

Visconti, G., M. Beniston, E.D. Iannorelli, D. Barba, Eds. (2001). *Global Change and Protected Areas*. Amsterdam: Kluver Academic Publishers.

Wanner, H., D. Gyalistras, J. Luterbacher, R. Rickli, E. Salvisberg, C. Schmutz, Eds. (2000). *Klimawandel im Schweizer Alpenraum*. Zuerich: vdf Hochschulverlag.

Weyant, J.P. (2000). *An Introduction to the Economics of Climate Change Policy*. Washinton, DC: Pew Center on Global Climate Change.

World Commission on Economic Development (WCED) (1987). *Our Common Future*. Oxford: Oxford University Press.

PART II

Impact

4. Economic Consequences of Climate Change in Alpine Regions: Impact and Mitigation

Walter J. Ammann and Veronika Stoeckli

1 INTRODUCTION

According to CIPRA (1998) about 11 million people are living in the European Alps, among them about 1.7 million in Switzerland. Snow, glaciers and permafrost dominate the area above 2 500 m a.s.l. Between 1 500 and 2 500 m, the snow cover still predominates over vegetation and soil for more than half of the year, and is thus an important ecological factor interacting with the flora, the fauna and the soil system. As an economic factor, the snow cover is most important in ensuring a prosperous winter tourism industry. Snow also influences the safety on roads and railways and is an important water resource. A rise in temperature due to climate change will raise the temporary snow line. This will have a direct impact on snow sports and thus on the economic situation of winter sport resorts at lower altitudes. Artificial snow may provide partial relief but it is unlikely to be fully appropriate as an alternative. An increase in temperature will also raise the equilibrium line altitude of the glaciers with consequences for the annual water balance, and will probably cause permafrost terrain to thaw, leading to potential slope instability. Besides changes in the abiotic environment, relevant changes in plant growth and distribution may occur. The influence of climate change on natural hazards is discussed in the final section of the chapter.

2 FORECASTING CLIMATE CHANGE IMPACT FOR THE EUROPEAN ALPS

With regard to the economic consequences of climate change in Alpine regions, it is essential that the probable extent of temperature rise and the influence on precipitation is discussed. The Third Assessment Report of Working Group I of the Intergovernmental Panel on Climate Change (IPCC WG I, 2001) builds upon past assessments and incorporates new results from the past five years of research on climate change. The global circulation models in use for climate scenarios are based on a relatively coarse horizontal resolution. Consequently, in order to gauge regional climate change, the simulated global scenarios have to be extrapolated to derive regional impacts; they have to be 'regionalised' by means of statistical methods and correlations to weather measurements, taking the large year-to-year variability into account.

The variability of Europe's climate is primarily determined by two pressure oscillation systems (Wanner et al., 1998): The one over the North Atlantic consists of the Icelandic Low in the north and the Azores High in the south (North Atlantic Oscillation), and the other one between the North Atlantic and the European continent essentially consisting of the Great Britain Low and the Urals High (East European Oscillation). These two large-scale alternations of atmospheric mass are dependent on the state of the ocean, and thus on the Gulf Stream, particularly over the long term where oscillations with periodical lengths ranging from years to even hundreds of years can be observed.

Significant fluctuations in the North Atlantic Oscillation have been noticeable in continental weather systems during the last 100 years, i.e. alternation between phases with pronounced west winds (moister in Northern Europe and drier in the Mediterranean region) and phases with pronounced north winds producing the opposite moisture conditions over Europe, the periodical lengths of these fluctuations amounting to ca. 20–30 years. In the Alpine region a cooler, severe winter period that began around 1950 was superseded in 1974 by a series of warmer years with more high-pressure systems in winter and generally fewer incursions of arctic air masses. This lasted until 1995. Since then the increased surface temperatures of the North Atlantic again indicate an increased incidence of invading cold air masses accompanied by more severe winters, a scenario which might apply for the next few years to come (Wanner et al., 1997). In any case, the Alps with 800 km in length, 200 km in width and a mean elevation of 2 500 m represent a formidable obstacle for invading air masses. This makes the elaboration of future climatic scenarios even more complex.

Due to the complex interaction between the atmosphere and the ocean, it is presently difficult to make any accurate predictions concerning a shift in seasonal weather system frequencies. Estimates of confidence in observed and projected changes in extreme weather and climate events during the 21st century are given in IPCC (2001): The globally averaged surface temperature is projected to increase by 1.4° to 5.8°C over the period 1990 to 2100. For Switzerland, the scenarios for an averaged air temperature increase vary between 0.8° and 3° to 8°C; they show a pronounced increase in winter precipitations (up to + 30%), a continuing retreat of glaciers and an increase in weather extremes, i.e. severe precipitations (ProClim, 2001).

These partially drastic changes in the Alpine regions may have a potential impact on the annual water balance, on hydroelectric power production, on summer and especially winter tourism, on mobility, on agriculture and forestry and on natural hazards. In the following section – due to the uncertainties in the scenario predictions – the evaluation of their economic impact can only be a rough and sometimes only a qualitative estimate. In addition, recent studies for Switzerland (Wanner and Schmutz, 2000, Gyalistras, 2000) point out that the scenarios – especially on a small-scale level – have to be evaluated with great care. Even when local trends were analysed using a variety of regional models, it was still not possible to eradicate controversy with respect to predicted trends. A similar lack of clarity remained when the Bavarian region was analysed. (BayFORKLIM, 1999).

3 IMPACT ON THE WATER BALANCE

The Alps are the water tower of Europe and as such of extremely high importance for the European lowlands. Looking at the Swiss Alps as an example, the Swiss Alpine section of the Rhine River to Basel covers only 23% of the total catchment area of 160 000 km^2, but contributes on average 50% or 1 000 m^3/s of the total discharge to the North Sea. The seasonal discharge varies from 30% in winter to 70% in summer, when stream flow is minimal in the lowland rivers, but high in the Alpine rivers due to snow and ice melt (Mountain Agenda, 1998). Similarly, large discharge contributions are observed in the Rhone River (32% of the Alpine area contributes 47% to the lowland flow) and the River Po (32% of the Alpine area contributes 56% to the lowland flow). During the summertime or the dry season, melt-water is released at a time when precipitation is at a minimum, and water demand is at a maximum. Switzerland thus has a great responsibility to ensure the passage of high-quality water to the countries downstream.

The importance of Alpine regions with regard to water resources is primarily based on enhanced precipitation due to orographic effects and resulting in much higher annual mean discharge rates per area (more than 30 l/s per km^2) compared to catchments in the Central European lowlands (about 10 l/s per km^2). On average, the annual precipitation in the Alps is 1 460 mm with an evaporation rate of 480 mm leading to a total runoff of 980 mm. A large portion of the precipitation falls as snow in higher altitudes and may be temporarily stored in the snow cover or in glaciers. The precipitation that reaches the ground, liquid or solid, is partitioned over time between evaporation, saturated and unsaturated subsurface flow, and overland flow, the relative magnitude of each component being largely determined by the nature of the particular surface, its gradient and soil depth.

A future temperature rise as shown in the scenarios above may significantly change the proportions of liquid to solid precipitations, and as a consequence the seasonal storage capacity of the snow cover and the long-term storage capacity of glaciers. Changes in average conditions (e.g. snow cover depth and duration, timing of the snow melt season, freeze thaw regimes or glacier extent) will lead to seasonal runoff changes with pronounced consequences in the river water regime in the lowlands. Although ice and snow in tropical, subtropical and mid-latitude mountains store only 1% of the global freshwater resources (Mountain Agenda, 1998), this supply is vital because of its proximity to the area of water demand.

An increase of the winter precipitations, as also stated in the scenarios above, might have contradictory effects. An increase in solid winter precipitation might have positive effects on the seasonal storage capacity and on the equilibrium line altitude of the glaciers. A rise in temperatures on the other hand will lead to earlier snow melt. Further research is needed to throw light on the various interaction processes.

4 IMPACT ON GLACIERS AND PERMAFROST

Glacier fluctuations in Alpine areas result from changes in mass and energy balance at the earth's surface. Glacier changes are among the clearest signals of ongoing warming in nature. Glacier recession, which has been visible in the Alps since the Little Ice Age (1700–1850 AD), will be even more pronounced in the future. In parallel with this, we are witnessing a rise of the equilibrium line altitude, i.e. the climatically determined glacier altitude where winter-spring snow accumulation is balanced by summer ablation. In Switzerland, the glaciers have lost about one-third of their volume since 1850 (Maisch et al., 1998), corresponding to a total water equivalent of 720 mm, i.e. 5 mm loss per year (taking the whole of Switzerland as reference area). During the same period, the glacierisation of the Swiss Alps lost up to 40% in surface area (Haeberli et al., 1999). At present, the Swiss glaciers cover a surface of 1 300 km^2 which corresponds to 3.1% of the total surface area of Switzerland. The equilibrium line altitude has risen 70 m (with an extremely high variation between – 115 m and + 445 m (Maisch, 1999)). During the last 20 years Swiss glaciers have lost about 300 mm in thickness every year, and taking the specific ice melting energy of 334 J/g into account, in order to melt, the ice needs a constant yearly energy flux of 3 Watt/m^2, which is quite consistent with the estimated anthropogenic greenhouse influence. With further temperature rise, the ice mass of Alpine glaciers could be reduced to only 25% of that found in 1850 within the next 30 years (Haeberli et al., 1999). Further examples on a global scale are given in Price and Barry (1997). Major economic impacts are to be expected: substantially reduced mean summer runoff will require new strategies in the seasonal management of the storage lakes for hydroelectric power generation (see Figure 4.2). In addition, the disappearance of a significant fraction of glaciers could have implications for summer tourism, by shifting the locations visited by mountaineers and other tourists.

There may be other less obvious, but nevertheless important consequences as a result of the gradual melting of frozen ground in permafrost areas. In Alpine regions, permafrost is widely present above 2 500 m a.s.l. Discontinuous permafrost tends to be highly sensitive. A temperature rise may force permafrost soil to melt, and potentially lead to a variety of slope instabilities in bedrock and non-consolidated sediments. An overview with regard to process understanding is given in (Haeberli et al., 1997). Mitigating risks due to such slope instabilities in Alpine areas requires systematic observation and monitoring of actual critical situations and an improved understanding of processes at work. Technical constructions built on permafrost ground may be exposed to accelerated creeping and settlement, if temperatures should rise further.

5 IMPACT ON THE SNOW COVER

During the last decade, the influence of climatic change on the snow cover has been broadly discussed. Of vital interest for tourist areas are the spatial and temporal development of the snow cover and the altitude of the snow line. The first and most important aspect is the proportion of solid to total precipitation, which depends on the altitude and regional climatic conditions. For Switzerland, Sevruk (1985) estimate such a proportion to be 3% of monthly snow fall per 100 m rise in altitude at 0°C. The freshly fallen snow in relation to the total precipitation at the test site of the Swiss Federal Institute for Snow and Avalanche Research

SLF in Davos-Weissfluhjoch at 2 540 m a.s.l. shows a total amount of solid precipitation of 75% which fits well with Sevruk's approximation (SLF, 1936–1998). A temperature rise will increase the proportion of liquid precipitation, leading to less storage volume and to faster runoff during winter time.

The seasonal snow height at a given site is very variable from year to year. The same is true for different regions over the Alps. Figure 4.1 shows the mean snow height over 10 years at seven different sites of similar altitudes all over Switzerland. Remarkable differences exist.

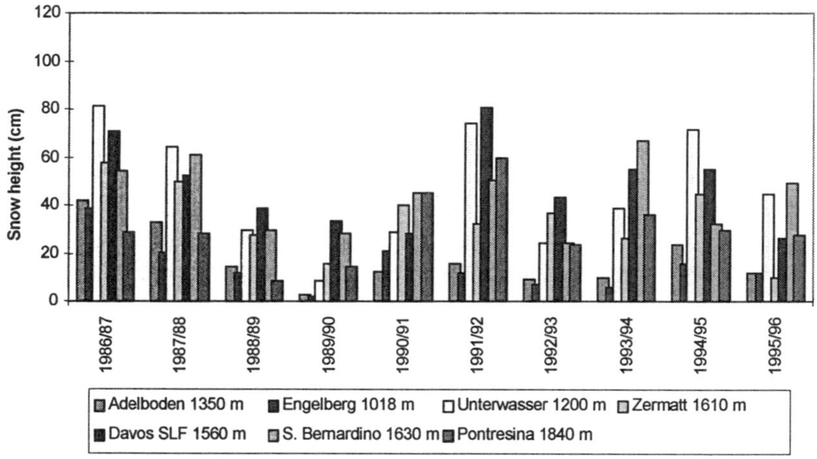

Figure 4.1: Mean snow height during winter 1986/87 until 1995/96 for 7 sites in different regions of Switzerland at similar elevations between 1 000 m–1 850 m a.s.l.

Snow is always very close to its melting point, therefore even minor temperature changes may have a dramatic impact on its presence and on its quality. An in-depth understanding of the various processes (precipitation as snow fall, snow accumulation, snow cover metamorphism) is necessary for adequate estimations of future changes. Foehn (1990) and Foehn (1992), in the discussion on the influence of a temperature rise of 3°C on snow cover development, concluded that:

- The altitudinal snow fall limit would rise 200–400 m.
- The mean altitudinal snow line would rise 300–500 m with the result that below 1 200 m a.s.l. closed snow cover throughout the winter season would no longer exist.

- At 1 500 m a.s.l. continuous snow fall would start later in autumn and the snow melt process would start much earlier in spring, leading to a shortening of the winter season by one month. The mean snow height would be reduced by at least 50%.

Recent findings by Schneebeli et al. (1998) showed no significant trends for the seasonal snow cover at the three sites Davos (1 560 m a.s.l., since 1898), Bever (1 710 m a.s.l., since 1910) and Andermatt (1 440 m a.s.l., 1947–1993). One of the reasons might be that the mean temperature during winter season is still well below 0°C for all three sites. In addition, temperature is only one of the major factors that have to be taken into account. General weather conditions are likely to play a more prominent role. As an example, the extreme avalanche winter of 1998/99 was not dominated by low temperatures but by several stable weather periods, all of the same meteorological configuration and leading to extraordinary snowfalls. Three precipitation periods accompanied by stormy northwesterly winds brought large amounts of snow to the Swiss Alps between 27 January and 25 February (SLF, 2000).

6 IMPACT ON WINTER TOURISM

Winter tourism relies highly on favourable snow conditions. Snow-deficient winters considerably affected winter tourism in the past. Climate change implications are therefore a new challenge for winter tourism. Snow reliability in winter sport resorts is becoming a major concern. Snow reliability was first defined by Witmer (1986), adopted by Abegg (1996) and finally adjusted by Buerki (2000). A ski resort is said to be snow-reliable if during at least 100 days between the 1 December–15 April the snow cover has a thickness of 30–50 cm in 7 of 10 winters. Abegg (1996) has analysed the snow reliability of all Swiss ski resorts. Table 4.1 shows the corresponding results, modified by Buerki (2000). For Switzerland today, snow reliability is guaranteed at an altitude of approximately 1 200 m a.s.l. Only 85% of the Swiss ski resorts are thus snow-reliable. If in the future the snow reliability could only be guaranteed at an altitude of 1 500 m a.s.l., the total number of Swiss ski esorts would be diminished by 37%, at 1 800 m a.s.l. by 56% and only 101 ski resorts would remain as snow-reliable.

According to the above definition, today about 85% of the Swiss ski resorts are still snow-reliable. A temperature increase of 3°C would raise the guaranteed snow reliability level up to 1 800 m a.s.l., resulting in a drastic reduction of skiable resorts. The economic consequences will be dramatic if no adaptation strategies are followed. Taking Switzerland as an example, we see that winter tourism has a value added of 11.5 billion SFr. corresponding to 1.5% of GDP. With more than 11 000 employees, winter tourism is the most important job provider in Swiss Alpine areas. Many farmers in these areas gain an additional income in winter time as employees in ski resorts. Swiss ski resorts had a total of 2 370 lifts in winter 1997/98, generating a gross turnover of 832 million SFr., 84% of which resulting from winter activities. In winters with poor snow conditions (1987–1989), the gross turnover was reduced by 20%.

Table 4.1. Snow reliability of the Swiss ski resorts for different altitudes and regions

Region	Number of ski resorts	Snow reliability at 1 200 m a.s.l.		Snow reliability at 1 500 m a.s.l.		Snow reliability at 1 800 m a.s.l.	
		absolute	%	absolute	%	absolute	%
Jura	15	4	27	1	7	0	0
Alpes VD + FR[1]	19	16	84	7	37	4	21
Valais	54	54	100	52	96	40	74
Berne	35	30	86	20	57	12	34
Central Switzerland	35	26	74	13	37	7	20
Ticino	8	8	100	3	38	2	25
Eastern Switzerland	18	11	61	6	33	3	17
Grisons	46	46	100	42	91	33	72
Switzerland	230	195	85	144	63	101	44

Note: [1]Alpes VD + FR: Canton Vaud and Canton Fribourg.

Source: Abegg (1996) modified by Buerki (2000)

About 300 ski lifts start below 1 000 m a.s.l. These stations may – even today – only be skiable for a few days per year. Most of the lift companies have a low cash-flow (less than 5%). More than 40% have a gross turnover of less then 100 000 SFr. Detailed information on the lift companies, their gross production and cash-flow may be found in Ammann (1999), SBS (2000) and BfS (2000). Especially the lowland ski resorts attract a lot of people from nearby densely populated areas. Many local children make their first attempts to ski in these modest resorts. If these places are no longer skiable in the future, ski sports may decrease in popularity leading to a decrease in winter tourism activities. The young generation, in particular, could then lose contact with winter sports and be lost as a future potential source of income in winter sports resorts (Buerki, 2000). A study by Meier (1998) puts expected yearly losses due to climate change at 1.8–2.3 billion SFr. (1.1–1.4 billion ECU) for Swiss winter tourism in 2050. Implications of climate change on winter tourism have also been evaluated in Austria (Breiling and Charamza, 1994). Similar conclusions regarding altitudes and snow reliability were found.

Adaptation strategies with regard to climate change are not easy to implement. The winter tourism industries focus heavily on short-term profit-making. Ski resorts are interested in knowing what will happen in the next 5 to 10 years and tend to be unwilling to adopt the lengthy time frame required. Nevertheless, due to the poor winter seasons in the late 1980s, a reaction has begun to take place.

Recent statistics show, for example, that during the last 10 years the number of lift facilities has been reduced by 6%, mainly in lower elevations up to 1 000 m a.s.l. The transport capacity has increased by 3% at the same time. This is mainly due to the replacement of older facilities by more powerful systems. Totally new constructions are focused on elevations between 2 000–3 000 m a.s.l. The general shift towards more snow-reliable elevations is remarkable (FOSD, 2000).

Another adaptation strategy is the production of artificial snow. Artificial snow is used for various reasons. One of the most important ones is to guarantee snow sports on downhill slopes from intermediate ski stations down to the valleys. It is also used to ensure a sufficient snow height for ski-slope preparation in early winter in altitudes up to 3 200 m a.s.l. Artificial snow production in ski resorts has been increasing continuously over the last 10 years. Today, the mean altitude of artificial ski slopes is around 2 000 m a.s.l., indicating that the present installations for snow production are mainly required not due to lack of snow, but rather to aid in ski-slope preparation. Early applications however were focused on lowlands and date back to the early 1970s (Ammann, 1999). The number of ski-slopes prepared with artificial snow differs between countries in the Alps. In 1999, Switzerland had a proportion of about 5% while Austria had about 20% (Ammann, 1999, CIPRA, 1998). Applications overseas indicate up to 30% in specific ski resorts in Australia (Buerki, 2000). The investment costs for the technical equipment are about 1 million SFr./km, and the seasonal service costs might vary between SFr. 30–50 000/km.

Changes in winter snow conditions are expected to have an ecological impact. This raises a host of new questions. An extensive, ongoing study by the Swiss Federal Institute for Snow and Avalanche Research, Davos, aims to provide answers. The focus of the study is the analysis of physical and chemical changes in ski-slope snow cover and how these changes interact with and impact on local vegetation and soil. In addition, the use of additives in artificial snow production and/or in ski-slope preparation, and their potential impact on fragile Alpine environments is also the subject of the analysis. Results have been summarised in Stoeckli and Rixen (2000). The vegetation composition on and beside artificially snowed ski slopes varies significantly. The treatment of artificial snow seems to favour late flowering species and snow-bank communities and to hinder early flowering species and wind-edge communities. This effect is attributed to a considerably postponed snow melt and start of the growing season on the slopes due to the additional mass of artificial snow.

The strong dependency on snow reliability is unfavourable for winter tourism. Alternatives to traditional winter tourism activities have to be developed. This is a long-term goal and might be the only way for ski resorts affected by climate change. In addition, this policy might also help to better protect the ecologically fragile Alpine environment. The first attempts in this direction have been described in Buerki (2000). The extent to which climate change is likely to affect tourism will depend, among other things, on the degree to which tourism has influenced the socio-political environment and the economy of an Alpine community.

7 IMPACT ON MOBILITY

According to CIPRA (1998), the regions defined by the European Alpine Convention have a total of more than 130 000 km of railways and roads of different categories. About 90 billion driven vehicle-kilometres result per year. Winter road service, i.e. snow removal and de-icing of the roads, may be very expensive and varies a lot depending on the road category, the location and the altitude. Examples of service cost are given for Switzerland in Ammann (1999). They vary between 2 000–3 000 ECU/km per winter in lowlands (up to 500 m.a.s.l.) and 12 000–16 000 ECU/km in high altitude. A mean value is not known but is estimated at 6 500 ECU/km. Exposed Alpine roads do not only have to be cleared of snow. Additional measures have to be taken to protect drivers against avalanches. For a pass road, service costs of up to 700 000 ECU per winter may thus occur. In extreme cases roads may even have to be closed during winter (Ammann, 1999). Most of the roads are kept snow-free using salt. For example, an amount of 10–20 tons/km of salt is used on the Swiss Federal Highways (Wieser, 1998). Winter service costs in lowland cities are over-proportionally high. The city of Zurich, for example, has paid an average of 2.5 million ECU per winter, with a mean total value of 51 cm fresh snowfall per year over the last 10 years. This sum corresponds to winter service costs of 50 000 ECU per cm fresh fallen snow.

Economic costs also occur due to speed restrictions on snowy or icy roads resulting in loss of productive working time. Estimations based on Ammann (1999) amount to a total sum for the European Alpine area of 2.5–4 billion ECU per winter. This sum does not include the costs of covering the increased number of road accidents during winter time.

An increase in temperature, as postulated by the climate change scenarios described above, with a rise of the mean altitudinal snow line and a reduction of the total snow height in lowland areas, will have a positive as well as a negative impact on the resulting economy. Winter service costs will decrease substantially and this might even compensate for the economic losses in winter tourism.

An additional point is worth mentioning. The catastrophic avalanche winter 1998/1999 showed that the avalanche risk for people on roads has been increasingly underestimated during the last decades. However, as analysed in SLF (2000), economically speaking, it is not justified to further reduce the avalanche risk by means of costly, permanent constructions. Assuming that climate change conditions will also lower the number of avalanche-prone days in higher regions, the overall risk will tend to decrease. The remaining risk may then be reduced by organisational measures. These temporary measures are much more cost-efficient. This requires an improved avalanche warning service and professional, overall risk management along the roads.

8 IMPACT ON HYDROELECTRIC POWER PRODUCTION

Hydroelectric power production is of vital importance for Austria and Switzerland. Relative to the total electric power production, hydroelectricity reaches up to 70% in Austria and 60% in Switzerland. The corresponding fractions for Slovenia are 30%, for Italy 20% and for France 15%. In absolute numbers, France occupies first place. The variability from year to year is substantial and rises to 10% (Chatelain, 1999).

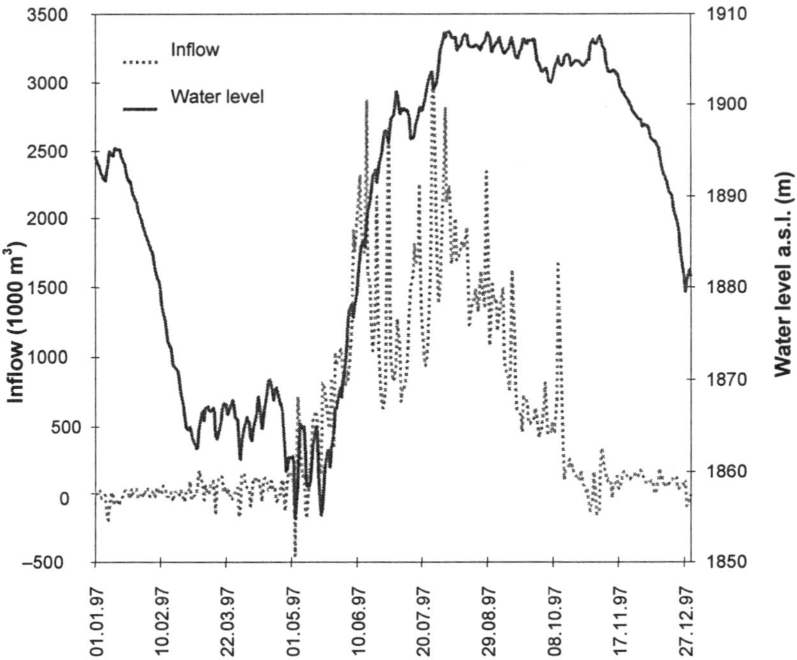

Source: H.J.Walther, Kraftwerke Oberhasli AG, Innertkirchen, 1998.

Figure 4.2. Hydroelectric reservoir of Lake Grimsel, Canton Berne: natural daily inflow (dotted line) and development of the resulting water level (solid line)

As described in sections 2 and 3, climate change will have various implications for snow melt and runoff. The management of reservoir lakes (see Figure 4.2), and hydroelectric power production along rivers, might need substantial changes in the future. The extent of change in time and intensity of the seasonal runoff as inflow to reservoir lakes will be crucial.

9 IMPACT ON AGRICULTURE AND FORESTRY

Relatively little research exists on the potential impact of climate change on Alpine agriculture. Predictions have been made that climate change will lead to higher climatic limits for cultivation. However, even if the climatic conditions are suitable for crops, the soil conditions may still limit growth. Soil characteristics tend to change much slower. Overall, crop growth is controlled by complex interactions between various abiotic parameters such as solar radiation, temperature, precipitation, drought, frost, wind, biotic agents such as pathogens or competitors and finally the methods of cultivation. In particular, the occurrence and effect of episodic short-term disturbances such as frost or drought is not clear. Only relatively little knowledge is presently available to help explain both the direction and degree of climate change impact on Alpine agriculture. The same is true for Alpine forestry. At present, it is only possible to speculate on long-term trends of mountain forest development, e.g. a general increase in growth and a rise of the upper altitudinal species and forest limits.

10 IMPACT ON NATURAL HAZARDS

The last decade was extensively prone to natural hazard catastrophes worldwide. Compared to earlier decades, the number of severe events increased by a factor of three, the economic losses even by a factor of more than 10. The year 1999 brought a total sum of economic losses of more than 100 billion US$, the insured losses rose to 20 billion US$ (Munich Re, 2000, see also Chapter 5). Around the world, extensive research programmes have been initiated in the past decade to investigate probable interdependencies between climate change and its influence on the occurrence and intensity of all different kinds of natural hazards. It was not often possible to deduce a link from observations in the past to climate change.

As an example, it was not possible to establish a trend for increased avalanche activity in Switzerland over the past century (Schneebeli et al., 1998). In February 1999, the Alps were hit by one of the most severe avalanche winters of the last century (SLF, 2000, Ammann, 2001). The statistically evaluated return period for this event was around 50 years. The last event of similar intensity occurred in winter 1950/51. The total economic loss at that time was only about one-sixth of the recent event the number of deaths however was six times higher (95 deaths in 1951 compared to 17 in 1999). Despite the lack of suitable evidence, the media and public discussions stated very quickly that the most likely reason for this severe event was due to climate change.

Similar findings are true for the Lothar winter storm event in late December 1999. A comparison with earlier winter storms in Switzerland showed that Lothar was well within the scope of what could be expected.

On average one storm with such high wind speeds sweeps across Switzerland once every 10 to 15 years (WSL, 2001).

Discussions on the amount of damage due to natural hazards and on the probable reasons behind them should always keep in mind the definition for risk. Already in 1978, UNDP defined risk as a mathematical product:

risk = hazard x vulnerability x value at risk

Hazard means the intensity of an event with a related probability of occurrence. This is the only factor that could be influenced by climate change. Climate change, with its influence on weather constellations, may have an effect on event occurrence intervals and on the intensity of natural hazard events. Vulnerability is expressed in terms of probable losses of life or goods, i.e. the amount of damage to buildings, infrastructure and ecosystems. The value at risk stands for the total amount of people (value of life) and of economic or ecological values exposed to a natural hazard event. 'Value at risk' is by far the most dominant element in explaining the extensive losses during the last few years. Again taking Switzerland as an example, the investments in buildings and infrastructure today are six times higher compared to 1970, and the number of cars is even 40 times higher compared to 1950 (SLF, 2000, Wilhelm et al., 2001).

Vulnerability also has to include all indirect losses resulting from subsequent damage (e.g. environmental pollution due to a pipeline broken in an earthquake, loss of market share due to damaged production facilities, etc.). Only a few figures are available that indicate the amount of indirect losses compared to the direct losses induced by the various natural hazard events. The reason is primarily the present lack of consistent and approved evaluation procedures. To gain more insight and to provide an example of evaluating indirect losses due to the avalanche winter and the storm Lothar in 1999, the Swiss Federal Institute for Snow and Avalanche Research SLF, Davos, conducted an inquiry on behalf of the cable-car and ski-lift companies in the German- and French-speaking part of Switzerland (Noethiger et al., 2001). In February 1999, avalanches and snow pressure damaged 44 cable cars and ski lifts, resulting in repair costs of more than 17 million SFr., which corresponds to 2% of the gross turnover of all companies. Due to high avalanche danger, 37% of all lift facilities had to be closed for an average of 7 days. The loss of earnings for Switzerland (excluding Ticino) is estimated at 78 million SFr. compared to February 1998. The winter storm Lothar caused damage to 127 cable-cars and ski-lifts. The cost of damage repairs was estimated at 7.6 million SFr. The storm interrupted power supplies to 14% of the lift facilities; 58% had to be closed down due to high wind speeds. The loss of earnings caused by winter storm Lothar amounted to 39 million SFr. (corresponds to 5% of gross turnover) for the Swiss cable-car and ski-lift companies (excluding Ticino). Immediately after the events of 1999, 32% of the companies interviewed took measures to reduce the negative consequences (e.g. price reductions, press releases or publicity campaigns).

Coping with risks due to natural hazards involves either influencing the value at risk or decreasing the vulnerability of people and goods so that the remaining risk is lowered to an acceptable level. A broad set of protection measures can be applied, depending on the type and intensity of natural hazard, its occurrence probability and the amount of people and goods to be protected. Generally speaking, the following protection measures can be applied in practice in order to prevent or at least reduce damage:

- Organisational measures (early warning, forecasting, alarm, temporary closure of traffic lines, evacuations of settlements, artificial hazard inducement)
- Hazard-mapping (as a base for land-use planning)
- Technical measures (all kinds of civil engineered construction work)
- Biological measures (silvicultural measures in protective forests, bio-engineered measures, etc.)

It may not be economically feasible to prevent a catastrophic event from occurring. In such a situation, an efficient intervention policy, i.e. immediate crisis management based on a previous emergency planning, is of vital importance in limiting further losses of lives and goods. But crisis management is only as effective as the reliability of the information on which the decisions taken are based. Thus, crisis management also means information management. The catastrophic events of the past two years have also revealed the importance of active information policy towards the public (SLF, 2000, WSL, 2001). The better people are informed, the easier it is for them to deal with difficult situations. It is also of importance that recovery measures are undertaken immediately after the crisis, so that those affected can resume normal life as soon as possible. A large amount of recovery cost is normally covered by insurance. Prevention, intervention and recovery thus form a unique set of measures. Also called 'integral risk management', it implies all the possible measures, which can be applied to reduce risks due to natural hazards, are of equal importance. Which set of measures will be the most effective in practice has to be determined based on cost-effectiveness calculations. Of course, it is not easy to compare prevention measures with intervention measures, or with an insurance strategy. An initial attempt was made by Wilhelm (1997), but it is in need of further development.

It is likely that future catastrophes due to natural hazards will continue to cause substantial loss of lives and goods. Damage statistics will continuously grow, as they did during the last decade. The main reason for this is the over-proportional growth of the 'value at risk', which is independent of any impact due to climate change. Climate change and its implications for natural hazard processes may be more pronounced in the future than current trend estimates reveal, since these have always been based on retrospective investigations.

11 CONCLUSIONS

Climate change involves a diversity of impacts whose economic implications are far from clear. In current scenarios, climate change is defined mainly in terms of temperature with only little account being taken of precipitation and probable weather extremes. This does not allow for in-depth estimates of the intensity of future events and, as a consequence, economic impact studies may only provide very rough estimates. Further research is necessary to improve knowledge on subsequent processes of climate change and to make the scenarios better applicable to the complex Alpine topography.

BIBLIOGRAPHY

Abegg, Bruno (1996), *Klimaaenderung und Tourismus. Klimafolgenforschung am Beispiel des Wintertourismus in den Schweizer Alpen*, Schlussbericht NFP 31, vdf Hochschulverlag an der ETH Zuerich.

Ammann, Walter J. (1999), 'Schnee und Lawinen: bestimmende Wirtschaftsfaktoren im Alpenraum', in Institut Universitaire Kurt Boesch (ed.), *Die Rolle des Wassers in der sozio-oekonomischen Entwicklung der Schweizer Alpen*, Institut Universitaire Kurt Boesch, Sion, pp.139–162.

Ammann, Walter J. (2001), *Avalanches – February 1999*, Swiss Report for EC Directorate General Joint Research Centre Ispra (http://nedies.jrc.it)

BayFORKLIM (1999), *Klimaaenderungen in Bayern und ihre Auswirkungen. Abschlussbericht des Bayerischen Klimaforschungsverbundes*, Bayerischer Klimaforschungsverbund, Universitaet Muenchen, Muenchen.

BfS (2000), *Statistisches Jahrbuch der Schweiz*, Swiss Federal Office for Statistics, Berne.

Breiling, Manfred and Charamza, P. (1994), 'Localizing the threats due to climate change in mountain environments', in Martin Beniston (ed.), *Mountain Environments in Changing Climates*, Routledge, London, pp.341–365.

Buerki, Rolf (2000), *Klimaaenderung und Anpassungsprozesse im Wintertourismus*, Ostschweizerische Geographische Gesellschaft, Neue Folge, Heft 6, St. Gallen.

Chatelain, Richard (1999), 'Die Nutzung der Wasserkraefte und deren Bedeutung im Alpenraum', in Institut Universitaire Kurt Boesch (ed.), *Die Rolle des Wassers in der sozio-oekonomischen Entwicklung der Schweizer Alpen*, Institut Universitaire Kurt Boesch, Sion, pp.91-108.

CIPRA (1998), *1. Alpenreport: Daten, Fakten, Probleme, Loesungsansaetze*, CIPRA International (Hrsg.), Verlag Paul Haupt, Bern, Stuttgart, Wien.

Foehn, Paul M.B. (1990), 'Schnee und Lawinen', in *Schnee, Eis und Wasser der Alpen in einer waermeren Atmosphaere*, Mitteilung Nr. 108, Versuchsanstalt fuer Wasserbau VAW, ETH Zurich.

Foehn, Paul M.B. (1992), 'Climatic change, snow-cover and avalanches', *CATENA, Supplement 22:* 11–21.

FOSD (2000), 'Entwicklung der touristischen Transportanlagen', Swiss Federal Office for Spatial Development, Berne (http://www.raumentwicklung.admin.ch/dossiers/dossier00_5_d.pdf).

Gyalistras Dimitros (2000), 'Zur zukuenftigen Entwicklung des Klimas im Alpenraum und in der Schweiz', in Heinz Wanner, D. Gyalistras, J. Luterbacher, R. Rickli, E. Salvisberg, Ch. Schmutz (eds), *Klimawandel im Schweizer Alpenraum*, vdf Hochschulverlag AG an der ETH Zuerich, pp.161–236.

Haeberli, Wilfried, M. Wegmann and D. Vonder Muehll (1997), 'Slope stability problems related to glacier shrinkage and permafrost degradation in the Alps', *Eclogae Geologicae Helvetiae*, Vol. 90, No. 3, 407–414.

Haeberli Wilfried, M. Hoelzle and M. Maisch (1999), 'Gletscher – Schluesselfaktoren der globalen Klimaaenderung', in José L. Lozan, H. Grassl and P. Hupfer (eds), *Warnsignal Klima. Wissenschaftliche Fakten*, GEO, 'Wissenschaftliche Auswertungen', Hamburg, pp.213–222.

IPCC WG I (2001), 'The Science of Climate Change', A Report of Working Group I of the Intergovernmental Panel on Climate Change, Third Assessment Report.

Maisch, Max, A. Wipf, B. Denneler, J. Battaglia and C. Benz (1998), *Die Gletscher der Schweizer Alpen. Gletscherhochstand 1850 – Aktuelle Vergletscherung – Gletscherschwundszenarien*, Schlussbericht NFP 31, vdf Hochschulverlag AG an der ETH Zuerich.

Maisch Max (1999), 'Vergletscherung und Gletscherschwund in den Schweizer Alpen aus glaziologischer, klimatologischer und hydrologischer Sicht', in Institut Universitaire Kurt Boesch (ed), *Die Rolle des Wassers in der sozio-oekonomischen Entwicklung der Schweizer Alpen,* Institut Universitaire Kurt Boesch, Sion, pp.109–137.

Meier, Ruedi (1998), *Soziooekonomische Aspekte von Klimaaenderungen und Naturkatastrophen in der Schweiz*, Schlussbericht NFP 31, vdf Hochschulverlag AG an der ETH Zurich.

Mountain Agenda (1998), *Mountains of the World: Water Towers for the 21st Century*, prepared for The Commission on Sustainable Development (CSD) and its 1998 Spring Session on Strategic Approaches to Freshwater Management, Mountain Agenda c/o Institute of Geography, University of Berne, Berne.

Munich Re (2000), *Topics. Jahresrueckblick Naturkatastrophen 1999,* Munich Re Group, 7.Jahrgang.

Noethiger, Christian, M. Bruendl and W.J. Ammann (2001), 'Die Auswirkungen der Naturereignisse 1999 auf die Bergbahn- und Skiliftunternehmen in der Schweiz', *AIEST 3-2001*, to be published.

Price, Martin F. and R.G. Barry (1997), 'Climate change', in Bruno Messerli and J.D. Ives (eds), *Mountains of the World. A Global Priority*, The Partenon Publishing Group, New York, London, pp.147–186.

ProClim (2001), 'Climate Facts', http://www.proclim.ch/ClimateFacts.html.

SBS (2000), *Annual Reports*, Swiss Cable Car Federation, Berne.

Schneebeli, Martin, M. Laternser, P. Foehn and W. J. Ammann (1998), *Wechselwirkungen zwischen Klima, Lawinen und technischen Massnahmen*, Schlussbericht NFP 31, vdf Hochschulverlag AG an der ETH Zuerich.

Sevruk, Bruno (1985), *Schneeanteil am Monatsniederschlag*, Beitraege zur Geologie der Schweiz, Hydrologie, vol. **31,** Bern

SLF (1936–1998), *Schnee und Lawinen in den Schweizer Alpen*, Winter Annual Reports, Swiss Federal Institute for Snow and Avalanche Research SLF, Davos.

SLF (2000), *Der Lawinenwinter 1999*, Eidg. Institut fuer Schnee- und Lawinenforschung SLF, Davos.

Stoeckli, Veronika B., Ch. Rixen (2000), 'Characteristics of Artificial Snow and its Effect on Vegetation', *Proceedings Int. Snow Science Workshop, Big Sky, USA.*

Wanner, Heinz, R. Rickli, E. Salvisberg, C. Schmutz and M. Schueepp (1997), 'Global Climate Change and Variability and its Influence on Alpine Climate – Concepts and Observations', *Theoretical and Applied Climatology*, 58.

Wanner, Heinz., D. Gyalistras, R. Rickli and E. Salvisberg (1998), *Die grossraeumige Klimavariabilitaet und ihr Einfluss auf den Alpenraum. Schlussbericht NFP 31*, Vdf-Verlag, Zurich.

Wanner, Heinz and Christoph Schmutz (2000), *Dynamische Ueberlegungen zur geographischen und synoptischen Kontrolle des Alpenklimas*, in Heinz Wanner, D. Gyalistras, J. Luterbacher, R. Rickli, E. Salvisberg and Ch. Schmutz (eds), *Klimawandel im Schweizer Alpenraum*, vdf Hochschulverlag AG an der ETH Zuerich, pp.31–58.

Wieser, Martin (1998), *Schweizerische Mittelwerte – Betrieblicher Unterhalt der Nationalstrassen*, Mueller AG Ingenieure, Betriebswirtschafter, Informatiker, Steinbockstr. 8, 7000 Chur.

Wilhelm, Christian (1997), *Wirtschaftlichkeit im Lawinenschutz. Methodik und Erhebungen zur Beurteilung von Schutzmassnahmen mittels qualitativer Risikoanalyse und oekonomischer Bewertung.* Mitteilungen des Eidg. Instituts fuer Schnee- und Lawinenforschung, Davos, Nr. 54.

Wilhelm, Christian, M. Bruendl, B. Brabec. S. Margreth and W.J. Ammann (2001), *Mobilitaet und Naturgefahren. Beitraege zu einem integralen Risikomanagement*, Proceedings 1st Swiss Transport Research Conference, STRC 2001, Monte Verita.

Witmer, Urs (1986), *Erfassung, Bearbeitung und Kartierung von Schneedaten in der Schweiz*, Geographica Bernensia, G25, Bern.

WSL (2001), *Lothar – Der Orkan 1999*, Eidg. Forschungsanstalt WSL und Bundesamt fuer Umwelt, Wald und Landschaft BUWAL (eds), Eidg. Forschungsanstalt WSL, Birmensdorf.

5. Climate Change and its Impact on the Insurance Industry

Thomas Loster and Martin Payer

1 INTRODUCTION

The big earthquakes in Turkey and Taiwan in 1999, in Gujarat in 2001, the landslide disaster in Venezuela (December 1999) and the typhoons (e.g. Taiwan/Nari 2001), hurricanes (e.g. USA/Allison 2001) and major winterstorms in central Europe (e.g. Anatol, Lothar, Martin at the end of 1999), provide further proof that natural disasters cause immense material loss and cost thousands of lives around the globe every year. For about twenty years now, the insurance industry has been worrying more and more about the rapid increase in claims burdens resulting from natural disasters. As a major proportion of these claims were caused by extreme atmospheric events, such as windstorms, floods, droughts and hailstorms, the suspicion soon arose that the environmental and climatic changes observed throughout the world could make a decisive contribution to disaster trends. Even though this relationship has yet to be confirmed fully scientifically, there can be no doubt as to the plausibility and, at the same time, the explosiveness of this assumption. Consequently, in accordance with the precautionary principle, industry and government must take the further aggravation of the disaster trend as a result of the anticipated climate changes into account in their considerations and draw up effective avoidance strategies to offset the costs.

In the sections below we first comment on the disaster trend in the last decades and its respective claims burden. Then the state of the art in research regarding climate change is presented, followed by possible scenarios for the 21st century including consequences for human life. Finally, the efforts to mitigate climate change on an international level, with specific reference to the insurance industry, are discussed.

2 DISASTER TRENDS

The claims burdens resulting from natural disasters have taken on dramatic dimensions, particularly in recent years and particularly for the insurance industry. Table 5.1 lists all the natural disasters in the past few decades that have cost the insurance industry more than US$ 1 billion (not cleared for inflation).

Before 1987, only one event, Hurricane Alicia in 1983, reached this figure. Since 1987, however, there have been 33 such events in total, 31 of which have occurred since 1990 alone. Hurricane Andrew still heads the list by a long way, causing insured losses in the region of US$ 17 billion, although the figure would have been several times higher if Andrew had achieved two direct hits in Miami and New Orleans, rather than a 'double miss'. The situation was very similar as regards the earthquake in California in 1994, which likewise only hit the edge of the Los Angeles district. Despite insured losses in excess of US$ 15 billion, this event can be regarded as just a 'warning shot', or a 'grazing shot' at most, like the Kobe earthquake in Japan in 1995. These two earthquakes are the only disasters in the list that were not of atmospheric origin.

The natural disaster trend since 1950 (Figure 5.1) very clearly reveals the dramatic increase in catastrophe losses in the last few years – a development that could in the near future result in annual claims burdens from major natural disasters in the region of US$ 25 to 50 billion. The decade comparison shows how big the increase was in recent years: in the last decade of the 20th century number of events, economic losses and the insured losses, between the 1960s and 1990s, increased by the factor of three, eight and fifteen, respectively (Table 5.2). These figures refer to so-called 'major' natural disasters. The remaining smaller natural loss events, of which the Munich Re records roughly 600–800 per year all over the world, increase the total claims volume to at least double this amount.

Without doubt, this increase in claims is largely, if not predominantly, caused by an increase in property values and insured liabilities, particularly in heavily exposed conurbation areas (Berz 1993). Moreover, natural disasters show time and again that the loss susceptibility of buildings and infrastructures has tended to rise, rather than fall, despite tighter building regulations and advanced technical developments. The observed upward trend cannot clearly be attributed to either climate or socioeconomic factors. The linkage between these two factors is complex and precise determination of the two causes – which themselves vary according to region and type of event – is hardly ever possible (IPCC 2001).[1]

Table 5.1: Natural disasters with insurance losses larger than 1 billion US$.

Rank	Year	Event	Area	Ins. losses (m. US$)	Economic losses (m. US$)
25	1983	Hurricane Alicia	USA	1 275	2 000
8	1987	Winterstorm	Western Europe	3 100	3 700
6	1989	Hurricane Hugo	Caribbean, USA	4 500	9 000
5	1990	Winterstorm Daria	Europe	5 100	6 800
24	1990	Winterstorm Herta	Europe	1 300	1 950
13	1990	Winterstorm Vivian	Europe	2 100	3 250
23	1990	Winterstorm Wiebke	Europe	1 300	2 250
4	1991	Typhoon Mireille	Japan	5 400	10 000
17	1991	Oakland fire	USA	1 750	2 000
1	1992	Hurricane Andrew	USA	17 000	30 000
18	1992	Hurricane Iniki	Hawaii	1 650	3 000
16	1993	Blizzard	USA	1 750	5 000
28	1993	Floods	USA	1 000	16 000
2	1994	Earthquake	USA	15 300	44 000
9	1995	Earthquake	Japan	3 000	100 000
27	1995	Hailstorm	USA	1 135	2 000
20	1995	Hurricane Luis	Caribbean	1 500	2 500
14	1995	Hurricane Opal	USA	2 100	3 000
19	1996	Hurricane Fran	USA	1 600	5 200
26	1998	Icestorm	Canada, USA	1 200	2 500
29	1998	Floods	China	1 000	30 000
22	1998	Hailstorm, Tempest	USA	1 350	1 800
7	1998	Hurricane Georges	Caribbean, USA	3 400	10 000
30	1999	Hailstorm	Australia	1 000	1 500
21	1999	Tornadoes	USA	1 485	2 000
12	1999	Hurricane Floyd	USA	2 200	4 500
10	1999	Typhoon Bart	Japan	3 000	5 000
15	1999	Winterstorm Anatol	Europe	1 800	2 200
3	1999	Winterstorm Lothar	Europe	5 850	11 000
11	1999	Winterstorm Martin	Europe	2 500	4 000

Source: Muenchner Rueckversicherungs-Gesellschaft (2001a).

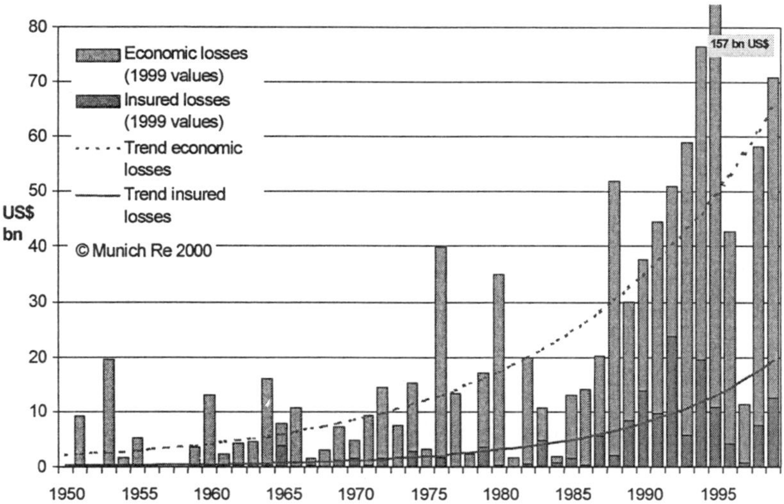

Source: Muenchener Rueckversicherungs-Gesellschaft (2001b, p.14).

Figure 5.1: Economic and insurance losses with trends 1950–1999

Table 5.2: Great natural disasters 1960–1999: decade comparison

	Decade 1960–69	Decade 1970–79	Decade 1980–89	Decade 1990–99	Factors 1990s:1960s
Number	27	47	63	89	3.1
Economic losses	73.1	131.5	204.2	629.2	8.1
Insured losses	7.0	12.0	25.5	118.8	14.9

Note: Loss figures in US$ billion (2000 values).

Source: Muenchener Rueckversicherungs-Gesellschaft (2001b, p.15).

At the same time, however, there are increasing signs that the emerging climate change is having an ever greater influence on the frequency and intensity of natural disasters. On the one hand, there are the major windstorm disasters of recent years that have set new loss records almost every year while, on the other hand, there are countless flood, tempest, drought and forest-fire disasters that today seem to be occurring more often than ever before.

3 CHANGES IN CLIMATE AND THE ENVIRONMENT

Since 1980 there has been one warm year after another, with the most dramatic conditions being recorded in the final years of the 20th century. The mean global temperature has reached its highest level since temperature readings were first taken world-wide almost 150 years ago. Since 1995 the nations of the world have organised various 'climate summits' in which they have engaged in negotiations on the various ways of at least mitigating or curbing the greenhouse effect. Although the coming decades alone will show how grave the effects of climate change are, the decisive question still remains: Can we really stop the hitherto uncontrolled experiment called 'climate change'?

Today a network of distinguished scientists spanning the world are investigating the greenhouse effect and the resulting climate change. The most reliable investigations appear to come from the Intergovernmental Panel on Climate Change (IPCC) that was set up by the United Nations. This is a panel of eminent climatologists from all over the world working with the support of thousands of other scientists, themselves sometimes producing very contradictory opinions. The 2001 published results of the IPCC's Third Assessment Report (TAR) make things crystal clear: Climate change is already taking place today.

Since the beginning of industrialisation in the late 18th century, there has been an accumulation of trace gases in the atmosphere. The main cause of this can only be the increased release of these gases by humankind. This applies in particular to carbon dioxide, which accounts for around 50% of the enhanced greenhouse effect caused by humans. In the course of time, the concentration of greenhouse gases has increased noticeably (e.g. CO_2: +30%, CH_4: +151%, CFCs: purely anthropogenic) owing to their mostly lengthy duration in the atmosphere. Today the first effects are clearly visible and scientific proof according to the IPCC's Third Assessment Report (2001) has now been established. There are too many alarming signs:

- The mean global temperature has risen in the past 100 years by 0.6 ±0.2°C, which may seem harmless at first. If we consider the fact that the differences between the ice ages and the moderate periods in the recent history of our climate have been no more than 5 degrees on average, however, the current change is remarkable.

- The 1990s have not only been the warmest decade, with 1998 being the warmest year since records began in 1861, but the temperature rise in the 20th century is also likely to have been the largest during the past 1 000 years.
- Even more evident is the rise in night-time daily minimum air temperature over land which increased by approximately 0.2°C per decade between 1950 and 1993. Compared to the rise in daytime daily maximum temperature (0.1°C per decade) this increase is twice as strong.
- As a consequence the freeze-free season in many regions has been lengthened too. According to satellite data this has been accompanied by a decrease of snow cover of about 10% since the late 1960s.
- An extensive and widespread mountain glacier retreat may be observed throughout the world and has already assumed dramatic proportions in some regions. The World Glaciers Monitoring Service, which looked at more a than thousand glaciers around the world, has come to the same conclusion. It was found upon investigation that the glaciers in the Alps have lost more than a third of their area and more than half of their volume since their last peak in the middle of the 19th century (Haeberli and Hoelzle 1995).
- The melting of the glaciers and the decrease in the extent of sea-ice contribute together with the thermal expansion of ocean water due to temperature increase to the rise in sea level. In global terms the rise of average sea level amounts to between 0.1 and 0.2 metres for the 20th century. The oceans are capable of storing large amounts of heat too and can thus act as a buffer. Nevertheless, there is proof that the surface temperatures of the oceans in the tropics have already increased. In the warm water pool of the West Pacific, which acts like a gigantic weather kitchen, significant temperature increases have been observed.
- Higher ocean temperatures lead to an exponentially higher rate of evaporation and to a correspondingly elevated water vapour content in the atmosphere. This promotes high precipitation intensities and tropical cyclones.
- Changes in temperature or precipitation patterns at certain times of the year do not only impact ecosystems. They can also trigger or intensify extreme events and natural catastrophes. In the late 1980s and early 1990s substantial windstorm damage was caused in Europe by a series of mild winters like those that are also to be expected in warmer climates. This was due to the fact that in winters with little snow, the cold high-pressure system over Eastern Europe is less

pronounced and thus some of its blocking effect is lost. In the end it allows the stormy low-pressure system coming in from the Atlantic to penetrate far into the continent. This was exactly the situation that was exhibited by the 1990 series of windstorms, which generated damage costing many billions.

- In many regions of the middle and high latitudes there has been a shift in rain distribution patterns both in regional and seasonal terms (e.g. in Central Europe where the frequency of hot dry summers and warm wet winters has increased).
- The frequency of heavy precipitation increased in the mid- and high latitudes of the Northern Hemisphere by between 2 and 4% over the latter half of the 20th century. In mild winters, when there is no snow, the rain fills the pores in the soil with water. Excess rain – from a winter low-pressure system for instance – cannot infiltrate the soil and is forced to flow off on the surface. The rapid swelling of flood waves in rivers then leads to flooding. Numerous flood catastrophes in Europe have already been caused by natural sealing, as this phenomenon is called, e.g. the ruinous Christmas floods on the Rhine and Mosel in 1993 and the Rhine floods of January 1995.

Many other changes have occurred. These include the hitherto inadequately addressed problem of large-scale destruction of the soil through deforestation, overgrazing, and other human activities, and the noticeable changes in fauna and flora – from declining biodiversity to the increase in tropical diseases outside their original areas. Climate change is just one of the many factors that are involved in these cases, but in many instances it plays a fundamental role. The IPCC (2001) concludes that '[t]here is new and stronger evidence that most of the warming observed over the last 50 years is attributed to human activities'.

3.1 What Can We Expect in the 21st Century? Scenarios on Climate Change

Do we need 'absolute security'? It will not be possible within the foreseeable future to forecast all the qualitative and quantitative changes and effects with complete precision. The most reliable statements, however, appear to come from the IPCC again. The decisive question is not whether or when we can expect definitive proof of anthropogenic climate change but whether the climate data already available and computer climate models can ever produce sufficient evidence on which to base a reasonable estimate of future changes so that the correct adaptation and avoidance strategies can be developed before it is too late.

The risk of error will continue to be sizeable in the foreseeable future but action must nevertheless be based on the best knowledge available at the current time.

In the IPCC's reports of 1990, 1996 and 2001 future development is extrapolated on the basis of a series of scenarios. This involves the analysis of time series to uncover trends and the use of highly sophisticated computer models. Before they are used, these models are subjected to a control experiment to test whether they are capable of reproducing the actual climate with the required degree of precision. Only then are plausible scenarios loaded and used as a basis for calculation. Here are some of the main results:

- In view of population development and the economic trends to be expected in the most densely populated regions of the world, it is likely that the emission of greenhouse gases will increase in the future at an ever faster rate. Thus, a doubling of the concentration of carbon dioxide even in the first half of the 21st century would appear to be quite realistic (pre-industrial: approx. 280 ppm, 1990: approx. 350 ppm, forecast for 2100: approx. 540–970 ppm; ppm = parts per million).
- According to the IPCC, the mean global temperature will rise by between 1.4 to 5.8°C over the period 1990 to 2100. The rate of warming that is responsible for this represents a massive acceleration compared with the observed fluctuations found per century over the last 10 000 years. The rate of change is so high that numerous ecosystems, including many types of forest, will probably be unable to adapt fast enough.
- The thermal inertia of the oceans mentioned above and the buffer effect of the increase in snow fall in the cold regions (e.g. in parts of the Antarctic) make it difficult to make exact forecasts. It is possible that sea levels will rise at a somewhat slower rate than originally feared. A rise by 0.09 to 0.88 metres is expected by the year 2100 (best guess is some 0.5 metres).
- The dramatic retreat of inland glaciers in the majority of the earth's mountain regions will continue. If the current rates of retreat are maintained, all that will be left of the once mighty Alpine glaciers at the end of the 21st century, for instance, will be a few meagre remains.
- An equally significant factor is the increase in humidity resulting from higher rates of evaporation, as this exerts a decisive influence on all precipitation and convection processes in the troposphere. Consequently, we will have to expect not only more torrential rain,

flash floods and debris flows, but also more thunderstorms, hailstorms, and lightning, as well as more tornadoes. As far as tropical cyclones – hurricanes and typhoons – are concerned, there will probably be an increase not only in their intensity but also in the length of their seasons and in their geographical coverage. Winter storms, i.e. extratropical stormy low-pressure systems, could also become stronger and could also, as already mentioned, penetrate further into continental zones.

- The results of recent research suggest that El Niño and La Niña may also be affected by climate change. Global warming could generate more frequent El Niño events or could intensify them or prolong them. These ENSO (El Niño-Southern Oscillation) phenomena, as they are called, have a lasting effect on the weather and on natural catastrophes in many regions of the world. Generally speaking, it seems very likely that the frequency and the extent of extreme weather and climate events will change during the 21st century.

3.2 Consequences of Climate Change Scenarios

The changes forecast above, despite remaining uncertainties, will also have enormous effects on the natural system and the socioeconomics of life. Both the natural system and the social and economic systems have a limited adaptive capacity. The vulnerability of regions to climate change varies greatly. Generally speaking, however, it can be stated that the less-developed countries in particular are most vulnerable. Their economies are based on climate-sensitive sectors and their ability to adapt to change, in terms of economic and human resources, is limited. The most prominent impacts, adaptations and vulnerabilities are again set out in the IPCC's TAR (2001). Some of the most relevant factors for Europe and the Alpine region are outlined below.

- Crop productivity may increase in Northern Europe due to an increase of only a few degrees.
- Half of the Alpine glaciers could disappear with the result that Europe will lose its largest drinking water reservoir. Further, inland shipping and the hydroelectric power plants will be affected.
- Higher temperatures affect tourism. Heat waves will influence the location of traditional summer destinations, and winter destinations will be subject to less reliable snow conditions.
- River flood hazards will increase dramatically through Europe.

- Extreme weather events and natural catastrophes will intensify and the weather variability will increase thus raising economic losses and the exposure of the insurance industry.
- The costs of fortifying infrastructure for climate change and an increased use of air conditioning during summer will compensate for or even outweigh the reduced costs of winter heating.

Table 5.3: Effects of global warming

Scientific certainty	
Very high	Increase in global mean temperature in the lower atmosphere and in upper oceanic layers Decrease in global mean temperature in stratosphere Temporarily severe ozone destruction in the polar stratosphere (ozone hole) Decrease in global ozone concentration in the stratosphere Melting and retreat of inland glaciers
High	Accelerated sea level rise Increase in atmospheric turbidity (aerosols) with regional cooling effects of the atmosphere Increase in frequency of mild winters with poor snow cover in Central Europe Increasing winter rainfall in Central Europe (decrease in Southern Europe) Increasing winterstorm activity over the Northern Atlantic
Low	Increasing winterstorm activity in Western and Central Europe Increasing tropical windstorm activity (frequency, intensity, source area, duration of storm season) Increasing activity of thunderstorms, torrential rainfalls and hailstorms in moderate climates Changes in fauna and flora Expansion of drought and desert zones in subtropical climates Spread of tropical diseases

Source: Muenchener Rueckversicherungs-Gesellschaft (1999, p.110).

The probability of antropogenic global warming is certainly high, but specific doubts remain (Lomborg 2001). Table 5.3 shows those effects that have already been observed or are expected, graded according to their degree of scientific certainty.

The costs and benefits of mitigation actions for climate change are currently the subject of a great deal of economic analyses. The results, however, still need to be handled with care. The reasons for these uncertainties can be found in the underlying assumptions on which the analyses are based. These include the rate and structure of demographic and economic growth, the occurrence of technological innovations, the level and timing of mitigation targets, the discount rates etc. Further difficulties arise regarding determination of the scope and the methodology of the analysis and the question of how welfare is measured (IPCC 2001). Apart from the prediction that the economic costs and benefits are not distributed evenly, neither among sectors nor among regions, general statements do not seem accurate particularly in the face of frequently changing national and international mitigation policies.

4 THE CLIMATE NEGOTIATIONS

It was not until 1992 that the starting orders were given for global endeavours specifically directed towards mitigating the human-made greenhouse effect, although the evidence of climate change had already been mounting before then. One of the most important aspects of the large-scale UN environmental conference in Rio de Janeiro was the agreement on a world-wide climate convention. It was affirmed that climate change could only be fought if globally binding agreements were installed that could also be monitored. In the initial years of the discussions, in which government representatives from more than 150 countries took part, it appeared impossible for the participants not to get bogged down in declarations of intent. Since Kyoto (1997), however, there is a sign of new hope. Binding programmes with specific target dates have been agreed upon: The main target is a reduction in global emissions of greenhouse gases by an average of about 5% of the 1990 levels by 2008 to 2012. In the 1998 meeting in Buenos Aires a detailed schedule on the operational issues of the Kyoto Protocol was defined with the critical deadline being the sixth global climate summit in The Hague, 2000. The negotiations in The Hague largely depended upon whether the major industrialised nations, i.e. those mainly responsible for causing the problem in the first place, ratify the protocols and thus accept obligations that are binding under international law. Although important negotiators like the EU were already quite definite in subscribing to the goals, other key countries, like the United States, refused to accept the terms as stated. The delegates agreed, however, despite their failure in reaching an agreement in The Hague, to resume their work in 2001 in Bonn.

After intense negotiations, the delegates at the conference in Bonn 2001 adopted a set of political compromises called the 'Bonn Agreement to the Kyoto Protocol'. This agreement fulfils the requirements of the Buenos Aires Plan of Action and is thus a further step towards a ratification of the Kyoto Protocol, mainly by ensuring a legal framework that serves to reduce the risk of the climatically relevant, long-run investments (Langrock 2001). In the long run this legal security can outweigh what is often called the 'bitter flavour of the agreement' (Ott 2001). By this is meant that the agreement cuts the greenhouse emissions reduction, set in Kyoto, by one half to about 2% of the 1990 level.

But speedy action is paramount, since even if emissions could be frozen to today's level, the CO_2 content of the atmosphere would still increase twofold in the next one hundred years. Besides negotiations at the highest level, activity 'from below' may also be promising, as shown by the Climate Alliance of European Cities, which developed out of the Agenda 21 process after the 1992 conference in Rio. It is not unusual for communal decision-makers to be involved in these working groups, and this often guarantees the implementation of measures. If the network of active politicians, representatives of the economy and citizens becomes tighter, the emission problem can be tackled quite effectively at its roots with the help of local core groups. After all, at least 80% of all the greenhouse gases that are released into the atmosphere are emitted in urban areas.

5 THE INSURANCE INDUSTRY – OPTIONS FOR ACTION

It is to be feared that in the wake of climate change in almost all regions of the world new extreme values will be recorded in many areas that have a bearing on insurance. This would imply natural catastrophes of unprecedented intensity and frequency. If this development is interpreted wrongly, the future of some lines of business or even the entire insurance sector could be at stake in some regions. That is why the insurance industry not only has a major stake in the climate negotiations, but also why it takes a proactive approach to protect itself from the impact of climate change and also to mitigate climate change itself.

The insurance industry and indeed the whole financial services industry are mainly regarded as a 'clean' business. Despite their green image, they have still been active participants in the Agenda 21 Process since the Conference in Rio 1992. In 1995, the most reputable insurance companies together with the United Nations Environmental Programme

(UNEP), launched the 'Statement of Environmental Commitment by the Insurance Industry' that was signed by 86 companies as of 1st November 2001 In this statement the signatories expressed their readiness to take more account of environmental risk in their core activities (VfU n.y.).

Starting from this statement these activities have become more and more active recently and are now bundled in the *UNEP-Financial Initiatives (UNEP-FI).* Various working groups have published papers and have been active in the field of risk assessment and prevention. The Climate Change Working Group – a part of the initiative – has held side events with statements at the most recent climate negotiations (conferences of the parties, COP 6b in Bonn and COP 7 in Marrakech, both in 2001). It has published a detailed position paper and has announced a comprehensive climate change study to be finished in 2002. Within this study – among other issues – appropriate market frameworks will be investigated that will encourage financial institutions to engage in adoption and mitigation actions.

Besides providing support in international research, fostering preventative actions and international co-operation, the insurance industry intends to play a role in 'defining solutions for managing climatic risks and contribute through risk management/insurance and through asset management' (UNEP 1997). As these two aspects concern the core business activities of the insurance industry they are discussed in more detail below.

Risk management

It is already certain that on the basis of the retrospective calculation of insurance rates commonly used at the present time, income from insurance premiums will constantly lag behind the development of claims. In order to protect itself efficiently against the consequences, therefore, the insurance industry must consider climate change in its price calculation. This, however, is not the case today. Yet, insurance industry is well-equipped to protect itself efficiently. What is more, it can make a substantial contribution towards the implementation of climate protection measures. This is because, like no other sector of the economy, the insurance industry has tools at its disposal that can generate the impetus for risk reduction. In managing the risk of climate change, however, all parties involved have to contribute. This means the insurance industry has to win both its clients and the authorities as its partners in sharing risk. For the insured, of course, this results in the bearing of a substantial deductible under natural hazards insurance, but on the other hand it produces some benefits for both parties to the contract *and* for the environment (UNEP 1997):

- The insured will be encouraged to adopt loss prevention or loss minimisation measures or otherwise bear a substantial deductible.
- The insurance industry can continue to provide its know-how in risk management and commercially insurable risk at a much more generous price.
- Insurers may be relieved of the bulk of minor losses, which can then be dealt with much more effectively by the clients themselves.
- While insurance companies are generally willing to offer cover, specific cases remain that are not commercially viable. In such cases the authorities responsible must be called upon to take measures that are designed to improve the risk or to fight the causes.

Asset management

Within the last ten years strategies for using the financial market to promote sustainable development have gained increasing attention (e.g. Schmidheiny 1992). The basic idea behind these concepts is the paradigm that environmental improvement and competitiveness of companies interact, thus that environmental improvement is good business (Porter and van der Linde 1995). Increasing numbers of companies are already concerning themselves with site analyses, eco-audits, and environmental reports, and are also examining their range of products in terms of environmental impact. These activities can be expected to be greatly intensified in the coming years and are said to be increasingly assessed by the stock market. As insurance companies, together with banks and investment funds, manage the majority of assets of global stock markets, using this investment style would have a great impact and would offer enormous leverage in the achievement of a sustainable economy.

Climate change and its consequences also lead to opportunities for innovative technologies especially in the field of regenerative energy. This all increases the need for a change in the use of conventional methods of company appraisal. More extensive appraisals that focus on the environmental risks and opportunities a company or industry faces have to be introduced. The existence of an increasing number of eco-rating companies like SAM in Switzerland, oekom and scoris in Germany, ARESE in France, Avanzi in Italy or Ethibel in Belgium shows the need for this kind of information. Such companies are engaged in assessing environmental (and social) activities of companies and deriving (environmental and social) information for investors and others, which supplements the traditional corporate information, which is mostly finance-based. This development has gained speed as a result of an increasing number of European government regulations require certain kinds of state-sponsored pension funds to disclose how, if at all, they take environmental issues in their investment policies into account (Mathieu 2000). The insurance industry as well as almost all the other members of the financial service sector, however, still have inadequate knowledge with regard to investment risk analysis and environmental investment strategies (UNEP 2000).

Further, the financial services branch including the insurance industry, will be able to use their know-how on financial instruments in the trading of emissions right, on verification of emission credits etc. (SAM 2001). These flexible market mechanisms as proposed in the Kyoto Protocol create new business opportunities (UNEP 2000) and will gain increasing importance on the world stock market. Furthermore, they present additional possibilities in diversifying portfolios of private as well as institutional investors.

These activities can be expected to be greatly intensified in the coming years and this will have a positive impact on protecting the atmosphere. If all the possibilities are used effectively, the consequences will quickly be very positive; i.e. protection of resources, improved environmental protection, image enhancement, and, not least, financial savings.

6 CONCLUSION

Humankind is in the process of performing a gigantic experiment with the earth's climate. As yet, it has hardly any control over this experiment and has no idea of the outcome. However, this experiment could have a dramatic effect on future conditions for human life on this planet. However possible it may still be to argue about the development of climate change and particularly about its effects, there are definite indications that the risk situation will deteriorate in the future. Efforts must be made to mitigate climate change and to restrict its impact as much as possible. The possible impacts for the insurance industry are dramatic, but the business is in the process of taking up this challenge in risk and asset management. Many steps, however, still have to be undertaken in order to reach a sustainable economy and thus to avoid even more dramatic impacts of climate change.

NOTE

1. Regression analysis, for example, is based on the time-constancy of parameters. The detection of structural changes is difficult, and can only be tested for single points in time, not for continuous or slowly growing changes.

BIBLIOGRAPHY

Berz, G. (1993), Global Warming and the Insurance Industry, *Interdisciplinary Science Reviews*, 18 (2), pp.120–125.

Haeberli, W. and Hoelzle, M. (1995), Application of inventory data for estimating characteristics of and regional climate-change effects on mountain glaciers: a pilot study with the European Alps, *Annals of Glaciology*, 21, pp.206–212.

International Panel on Climate Change (IPPC) (2001), *Third Assessment Report: Summaries for Policymakers.*

Langrock, T. (2001), Klimaschutz – Die Bonner Beschluesse, *Umweltwissenschaften und Schadstoff-Forschung*, Bd.13, Nr.4/2001.

Lomborg, B. (2001), *The Skeptical Environmentalist*, Cambridge University Press, Cambridge.

Mathieu, E. (2000), *Response of UK Pension Funds to the SRI Disclosure Regulation*, UK Social Investment Forum, London.

Muenchener Rueckversicherungs-Gesellschaft (1999), *Topics 2000 – Naturkatastrophen: Stand der Dinge*, Muenchen.

Muenchener Rueckversicherungs-Gesellschaft (2001a), *NatCatSERVICE-Elementarschaden-Datenbank* der Abteilung GeoRisikoForschung, Muenchen.

Muenchener Rueckversicherungs-Gesellschaft (2001b), *Topics Jahresrueckblick Naturkatastrophen 2000*, Muenchen.

Ott, H. (2001), The Bonn Agreement to the Kyoto Protocol – Paving the Way for Ratification, *International Environmental Agreements: Politics, Law and Economics*, Vol.1, No.4, pp.469–476.

Porter, M., C. van der Linde (1995), Green and Competitive: Ending the Stalemate, *Harvard Business Review*, Sept.–Oct. 1995, pp.120–134

Schmidheiny, S. (1992), *Kurswechsel*, Muenchen.

Sustainable Asset Management (SAM) (2001), *Klima-Abkommen in Bonn – Bestaetigung fuer Sustainability-Investments*, http://www.sam-group.ch/d/PDF/Inv_Info_Bonn_010724_d.pdf

United Nations Environmental Programme UNEP (1997), *UNEP FI Working Group: Climate Change and the Financial Sector – Position Paper presented at the Conference of the Parties of the United*

Nations Framework Convention on Climate Change, http://unepfi.net/cc/ccpp97.htm

United Nations Environmental Programme (UNEP) (2000), *UNEP Finance Initiatives: Climate Change Working Group Scoping Paper*, 1st edition, Geneve.

Verein fuer Umweltmanagement in Banken, Sparkassen und Versicherungen (VfU) (n.y.), *Time to Act – Environmental Management in Financial Institution*, Bonn.

PART III

Adaptation

6. Human Vulnerability – Factors Influencing the Implementation of Prevention and Protection Measures: An Agent-based Approach

Doerthe Kroemker and

Hans-Joachim Mosler

1 INTRODUCTION

The Alps belong to one of Europe's most beautiful regions. They provide recreation for thousands of tourists every year and are of considerable importance for the economic prosperity of the Alpine nations. But the Alps are also a source of danger. Avalanches, floods and landslides have always been part of the region. But they turn into a disaster as soon as humans are affected, that is where they lead to death of tourists and locals and cause damage to homes and belongings. Anthropogenic-caused climate change is expected to cause more precipitation and a higher frequency of extreme weather events such as thunderstorms, storms and heavy rainfall. Thus, such disasters are expected to occur more often in future (see Chapter 5). Certainly the impacts of such events are intensified by regional human activities, such as deforestation, decline of forests due to traffic-based air pollution or overgrazing, which lead among other things to a destabilisation of slopes, vegetation and soils. The presence of infrastructure and houses in potentially endangered regions further enhances the risk of disasters. Humans cause such environmental changes, but humans are also victims. In the Alps – as in many other regions of the world – a double vulnerability can be seen: the vulnerability of the ecosystem and the vulnerability of the human social system.

Human social systems and ecosystems are closely intertwined. This holds true for every scale of analysis: global, regional, local. The development of one system cannot be understood and changed without regard to the impact on the other one. Consequently, both the analysis of the causes of environmental changes and the derivation of intervention strategies that facilitate change towards more sustainability is an interdisciplinary task – and an enormous challenge. Nevertheless, in order to complement those articles in the book analysing the natural system, the present contribution focuses on aspects of the social system and does so from a psychological point of view. Further effort will most certainly be needed to bring these diverse approaches closer together in future.

The following section highlights those factors that may lead to an increased vulnerability of the human social system with respect to natural hazards. Though no simple answer can be given, a certain perspective – an agent-based model – is introduced to analyse the relationship between humans and nature. This general human-nature interaction model is helpful in explaining the difficulties in implementing suitable prevention and protection measures that are needed to minimise both, the social and the ecosystem vulnerability. The framework is based on psychological action and stress theories (Lazarus and Folkman, 1984; Rippetoe and Rogers, 1987; Schwartz and Howard, 1981). These have also been applied in environmental issues within theoretical and empirical research (for instance Gardner and Stern, 1996; Kannapin et al., 1998). But the comprehensive and general theoretical framework developed in this chapter is a new approach within vulnerability research, has not yet been tested empirically and has certainly not been implemented in the Alps. However, this general framework may hopefully offer a basis for the analysis of circumstances specific to the Alps as well as other issues related to vulnerability (compare also Kroemker and Mosler, 2000).

2 VULNERABILITY AND CRITICALITY

Before the agent-based model is introduced a few words about the term 'vulnerability' need to be said. Several definitions of vulnerability are currently in circulation. These are well covered in various publications and therefore need not be repeated here (for an overview see Cutter, 1996; Blaikie et al., 1994; Bohle et al., 1994).

There is no uniform understanding of the concept and no consistent theoretical basis has been developed up to now. However, roughly speaking, most vulnerability approaches have, despite variations in focus or analysis, the following two factors in common (Kelly and Adger, 1999):

- Criticality: this analyses stress, hazard or exposure to hazards (or in terms of risk: the probability of exposure to hazards) usually in relation to geographical regions
- Capacity: this analyses ability to cope, to adapt, to recover usually in relation to social, economic, historical and political conditions

Alpine regions show a high probability of exposure to hazards, with some areas being more and others being less risk prone. However, we assume that criticality is given in the Alps and are not interested in defining variations in hazard probability. The analysis in this chapter centres on variations in capacity levels and their relevance for human vulnerability. Hence we associate ourselves with the approach of Blaikie et al. (1994, p. 9), who define vulnerability as 'the characteristics of a person or group in terms of their capacity to anticipate, cope with, resist, and recover from the impact of a natural hazard.'

Using this logic, vulnerability is low, if capacity is high and vice versa. Capacity is understood in a very comprehensive way. It includes preventive strategies, that is those with the aim of minimising the probability of a hazard or the possible impacts of a hazard; it includes coping strategies, that is those used in a disaster situation and ability to recover fast after a disaster. A state of high capacity results from human deliberation and action. That is why we focus on agents, understood as persons or groups of persons, who deliberately interact with their surroundings – the physical and social environment.

Essentially, the main question in the context of vulnerability is then: Which factors lead to successful action strategies, which factors lead to unsuccessful action strategies in the face of critical environments?

3 PROTECTION CAPACITY

Protection capacity consist of the above mentioned factors such as anticipation, coping, resistance, recovery and – not to forget, prevention. It is not possible to judge in general which specific action strategies are needed in order to maximise protection capacity. In fact this is often a source of conflict as is known from risk research (for example Edwards and Winterfeld, 1985; Bonß, 1995). Protection of the ecosystem and of the human social system in the Alps results from omitting as well as undertaking actions. Not to build a new ski-run, not to build a new hotel and a new road, to give up very exposed villages or to reduce traffic might be just as suitable in generating high protection capacity as would be, for example, readopting traditional cultivation techniques, building protection fences or implementing other technical measures, training rescue teams for example. Whatever these strategies are in detail, when successful they help to prevent disasters or minimise their impacts. What is right is the subject of negotiation among local stakeholders. In the end, a lower number of casualties, less financial damage and a healthy environment indicate success.

We now turn to those psychological factors that influence participants willingness to enhance protection capacity, that is to implement prevention and protection measures.

4 MOTIVATION AND COMPETENCE

Put simply, only two factors influence the development of protection capacities: motivation and competence (see Figure 6.1, protection capacity schema for depicting the combination of the two factors). Persons who really want to do something to defend themselves or others from harm, but do not have the competence to put their wants into practice fail (Type B) as do persons who would be capable of putting measures into practice, but do not feel that there is any danger that would make them necessary (Type C). Persons of Box B keep seeking to do something but often show a 'coping style of avoidance': They choose strategies that cannot change the situation itself, but do bring back cognitive control (see section 6). Persons who are both motivated and able to install protection measures are the most likely to realise them (Type D), whereas persons who are neither motivated nor capable do nothing at all (Type A) (Gardner and Stern, 1996). Figure 6.1 indicates these four rather artificial ideal types, though the existence of types 'in between' is probable. Usually, all different types of persons are members of one social group (for example a village or a district) and share the same environment, but some think it is close to collapse (as typically nature conservationist groups do) and others do not even recognise a sign of disturbance (typically groups that want to see a region develop economically). The more homogeneous the realities of a social group are, the easier it is to realise any action or non-action. The more persons of the type 'motivated and capable' share a group, the more likely it is that protection measures are achieved. To repeat a point mentioned above: Whether the measures are the effective optimum or not cannot be discussed here.

Social groups are dynamic and types change over time. It is likely that persons move from one type-box to the other in the process of time due to e.g. new external events or internal changes. Beside this, new persons and therefore new motivation-competence-types can become members of the group whereas other persons can leave the group. All this is a source of social learning and innovation.

So far the basic level of interplay between motivation and competence has been presented. But it is also important in developing sustainability improving strategies to analyse which factors feed different degrees of motivation and competence. These factors are introduced in the next section.

	Appraisal of motivation: *low*	Appraisal of motivation: *high*
Appraisal of competence *low*	Type A: No protection capacity Lack of both, motivation and competence	Type B No protection capacity Lack of competence 'Avoiding' coping style is likely
Appraisal of competence *high*	Type C No protection capacity Lack of motivation	Type D Protection capacity is likely

Source: Adapted from Gardner and Stern (1996, p. 245).

Figure 6.1: The Protection Capacity Scheme

5 FACTORS INFLUENCING MOTIVATION AND COMPETENCE

5.1 Motivation

The interplay between the individual's or group's values, knowledge, aims, personality traits and the characteristics of the physical and social environment, provide a starting point for an initial appraisal of motivation in any given situation (see Figure 6.2 for an overview of the factors influencing motivation). Depending on the outcome, a person is more or less motivated to look for protection measures. If a person thinks the situation is completely irrelevant for her or others' well-being, she will not be motivated to think about the issue any longer. For that person other things of daily life will be more important. If, on the other hand, somebody thinks the situation is relevant, different appraisal outcomes are possible. To name the most important, this could be a sense of threat, of challenge, or responsibility, or some such mixture (Lazarus and Folkman 1984; Fuhrer and Woelfing, 1997).

A sense of threat, although very prominent and often fostered by the media ('climatic catastrophe'), is not the only driving force underlying the search for protection and prevention measures (see right side of Figure 6.2): Many of the recent disasters in the Alps are the product of long-term interventions in the landscape or of other long-term cumulative factors like traffic-induced air pollution or global warming. Here, the possibility of a direct and immediate harmful feedback is not given or not perceived and therefore a sense of threat often does not provide a source for the motivation to implement protective measures. But a person's ascription of responsibility and moral obligation can act as a driving force. In addition, those persons who do not feel personally threatened, but feel challenged or perceive that there is a danger for nature and animals or for future generations, can be motivated to implement protection measures. Of course this by no means rules out that those persons feeling directly threatened can also feel additional responsibility for preventing others coming to harm.

Under what conditions will a person accept the responsibility of providing protection? First, a person must think the situation is problematic and be aware of the consequences if no protection measures are implemented. In addition, the person must believe that her own individual actions are decisive. If this is not the case it is likely that others will be left to bear the responsibility. Besides this, it is found that responsible agents often exhibit strong personal norms (e.g. they say 'I should not use my car all that often, I should do more to protect the environment') and that they also have the impression that other persons expect them to protect the environment (Bloebaum et al. 1998; Guagnano et al., 1995).

Due to the nature of many environmental problems (i.e. those with non-perceivable, or rather long-term effects), it is important that society finds ways to foster and to promote responsibility, especially that of powerful agents who have a high impact on collective vulnerability (see section 5.2).

Many of the factors enumerated below allow persons to perceive a situation as irrelevant for well-being and therefore not worth worrying about. That may explain why the group of the non-motivated type seems to be bigger in many communities and may explain a part of the difficulties in implementing preventive protection measures. The factors described below are important in generating protection motivation, whether they arise from a perceived threat, a sense of challenge, or a feeling of responsibility.

Characteristics of the environment

Nature often gives no warning perceptible to human senses: many disturbances in the ecosystem do not smell, make loud noises or produce visible signs. What may be visible might simply be overlooked due to habituation. The lack of healthy forests and soils for example might not be visible to the untrained eye. In addition, information is often contradictory and uncertain – for example, people learn from the media about the impacts of climate change or the dying of forests, but are unable to relate it to their present experience and may ask 'Why should one worry, when the Alps are as beautiful as ever'. Consequently public pressure is often missing and therefore cannot support the drive for protection measures. Uncertainty about future environmental developments also makes it difficult for experts to judge which level of protection and prevention measures provide a necessary minimum.

Knowledge and characteristics of the social group

A certain environmental knowledge about cause-effect relations and possible future damage is necessary in order to perceive a potential danger. When such knowledge is not available, it is unlikely that persons are motivated to look for protection measures. Also biased probability assessments ('we had the flood last year, it cannot happen again this year'), which are well documented in risk research (for example Slovic et al., 1985), reduce the likelihood of generating a motivation for protection.

Knowledge, whether provided by the mass media, or through social interactions, for instance via face-to-face communications, helps to constitute a local reality about the issue at stake. Also for experts, knowledge, creation and interpretation, and judgement formation contain many uncertainties and can be influenced by belief systems, values and interests (Shrader-Frechette, 1995; Mayntz, 1996). Furthermore, these are often the product of specific constellations of power, as has been well documented in risk research (see Chapter 8). For many environmental issues, it holds true that members of different social groups, as for instance nature conservationists opposing a communal planning committee, each refer to different sets of knowledge and consequently come to different conclusions about the severity of risk (Renn, in press).

Hence, the quality of knowledge not only influences the strength of motivation behind the generation of protection measures, it also, as will be shown later, influences the choice of strategies when dealing with any specific situation.

Values and aims

Even where a person is aware of a potential danger, she must feel that important material or immaterial values are at stake. The term values can here encompass very different things like own life, health, home, social history, aesthetic nature, belongings or livelihood. The perception that important values are endangered depends – among other factors – on the specific situation agents are in. Different aims and claims with respect to landscape use play an important role. There might be a difference in the perception of the severity of potential damage between for example the manager of a big hotel company on the one hand and an owner of a small guesthouse on the other hand. The manager is perhaps not rooted in the area, whereas the guesthouse owners' family has been living there for generations. Besides this, the former might have the financial power to invest a considerable amount of money, while the latter might not have the resources to do so. For the manager it could be easier to assume the risk of changing the landscape by for example building a new ski-run. In the case of a devastating landslide, he would merely lose one job opportunity among many others; for the owner of the guesthouse, however, this could mean a complete loss of livelihood or – perhaps even worse of – his or her home. One other important difference is relevant here: the manager would assume the risk voluntarily, whereas the guesthouse owner would feel a victim, a situation that decreases the willingness to accept risk (e.g. Jungermann and Slovic, 1993).

Besides the specific aim derived from the socioeconomic context people live in, more general beliefs about what is good or bad are also important in terms of their effect on protection motivation: the hotel manager for example could be convinced that humans do not have the right to take more from nature than they already have. He could therefore argue against a new ski-run. The more people feel their central values of any kind are endangered, the more they are motivated to protect them.

Things can get even more complicated, since people usually have a whole bunch of important – often contradictory – values. Both the hotel manager and the guesthouse owner will probably have an interest in providing an attractive recreation area for potential winter guests, and therefore be strongly in favour of a new ski-run. Attractive new investments will only be avoided where perceived risks outweigh expected benefits. And developing this reality is a tough task, especially when there are no obvious signs of danger. And where danger does finally become visible for example rising rivers loaded with mud and stones, it is often too late.

Personality traits

The perception of the severity of damage can also be influenced by personal characteristics such as anxiety, trust in a good fate or positive or negative thinking. The entrepreneur, who perceives the situation rather as a challenge than as a threat, might be somebody who believes that things can be controlled due to good planning and high-quality security standards. This person is probably motivated to undertake measures that minimise the impact of a potential disaster, but measures that could prevent the occurrence of a disaster probably are perceived as exaggerated and as threatening economic prosperity. Likewise the person ready to ski down a side run could feel encouraged by the challenge of danger.

Figure 6.2: Factors influencing the appraisal of motivation

In summarising the above discussion, we can say that the specific combination and peculiarity of values, knowledge, aims, personality traits, characteristics of the environment and the social group lead to different subjective perceptions of the physical and social environment, which are shared with other specific individuals of the social group (for example a community). These perceptions lead to an appraisal of immediate threat, of responsibility or of challenge. Such appraisal process provides a source for the motivation behind the drive for establishing protection measures. There are also people who develop no sense of threat, responsibility or challenge. For such people the existing situation may have no relevance whatsoever as far as the need for protection is concerned.

5.2 Competence

The motivation to undertake protection measures of any kind is an important prerequisite for the development of protection capacity, but is by no means sufficient. The ability to implement preventive measures or to minimise hazard impacts is crucial as well. Similar to the processes needed in generating motivation, the appraisal of such competence is influenced by psychological factors like values, knowledge and personality traits. But here a multitude of situational barriers are highly relevant.

Before the several factors influencing the appraisal of competence are introduced in detail, two principle aspects have to be stressed:

1. Agents have different institutional entitlements and social roles.
2. These entitlements crucially affect the capability of persons in preventing collective or individual vulnerability.

In general, actions that potentially lead to low collective protection capacity or high collective vulnerability and actions that lead to low individual protection capacity or high individual vulnerability are distinguishable. The consequences of the first type of action potentially affect a larger group of people and are therefore more important for the reduction of vulnerability (for example as a result of more avalanches or landslides due to the destabilisation of slopes). Such actions require an organised group of decision makers (for example a parliament) for their realisation. However, actions that lead to collective vulnerability can also be the product of a larger group of unconnected non-organised agents (for example car-drivers whose cumulative actions cause the dying of forests) or a single person (for instance skiing down a side run thereby causing an avalanche and endangering other persons). In contrast, individual vulnerability is given when a person does not undertake actions to protect herself or himself or her or his close social reference group. A family that refuses to leave their house in an endangered mountain village or one that has no means of preparing the house for flooding is an example of this kind of vulnerability.

The institutional entitlement of a person obviously affects the quality and amount of possible action strategies she or he can implement and thus whether he or she can provide for individual or collective vulnerability. A potential investor for example can easily prevent a new ski-run by simply giving up the project. For members of the local authorities (for instance a local political board), influencing general conditions towards securing more environmental safety may be possible, but depending on the majority in the respective organisations this can be hard to achieve in practice. The 'normal citizen' or nature conservationist groups can try to influence and convince decision makers through a range of actions (for example establishing a pressure group, organising demonstrations, face-to-face communication and so on), but usually they do not have any direct mechanism of control and sanction at their disposal.

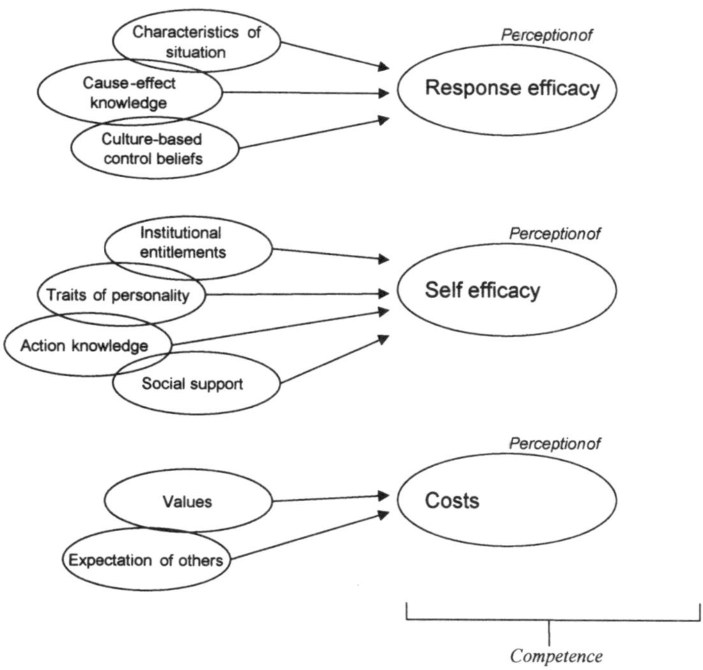

Figure 6.3: Factors influencing the appraisal of competence

While the institutional action opportunities do make a difference to the potential outcome, the prototypical core of the appraisal process, described below, is the same for all agents. Figure 6.3 depicts the factors relevant for the appraisal of competence.

Knowledge, cultural beliefs, characteristics of the situation
The person assesses which actions are available at all when responding to a situation (perceived response efficacy) that she or he might perceive as a threat, a challenge or one requiring responsibility, as introduced above. The more experience and knowledge (cause-effect knowledge) a person has, the more action possibilities she or he has to choose from. Those decision makers in a community considering technical solutions, afforestation and abandonment of specific villages have a higher response efficacy than those only counting on technical solutions. Personal characteristics such as mental flexibility or creativity tend to enlarge the room for action possibilities. Also culture-specific beliefs, for instance the conviction that one's fate is determined by God can restrict response efficacy or the belief that one is solely responsible for one's own fate, may enlarge the room for coping strategies (for example Tafarodi and Walters, 1999). The given external situation is also relevant. For example afforestation is no option where this is ecologically not feasible.

Personality traits, knowledge, social support, institutional entitlements
Not only the principal action opportunities are of relevance, but the person's perception of being able to perform the respective action is important as well (perceived self-efficacy). Internal and external barriers influence that personal ability. Examples of internal barriers are personality traits such as negative thinking and negative control beliefs ('Nobody will listen to me', 'I cannot do anything on my own'). Knowledge in the sense of know-how (action knowledge) is of relevance again. For example, a citizen could know that establishing a pressure group for organising public attention could be an action opportunity to prevent the realisation of a new hotel, but has no idea at all how to start one.

Examples of external constraints are the above-mentioned institutional entitlements or a lack of financial or other material means. The more possibilities people feel they have, the more likely they are to assess their competence as high.

Values, expectations of others

A final and very important factor for the appraisal of competence is the evaluation of costs for possible protection measures. Costs are here understood in a very comprehensive way: besides the financial aspects, also effort, time, not being in accordance with personal norms, getting in conflict with other important values in live, not following social norms and so on, all produce costs for agents. The agent assesses whether the possible action strategies conflict with other important issues. For example the mayor of a village may feel responsible (motivation) to prevent the granting of building permission for new hotels in an area exposed to risk. She or he also feels competent to undertake the necessary steps (for example convincing other people). But she or he also thinks that people expect her or him as a politician to care for local economic and social prosperity. If the costs of not following such expectations are higher than of not following her or his feelings of responsibility, the mayor will promote the hotels. Also the preference for short-term technical solutions over long-term security projects (for instance afforestation) as reported by Weiss (see Chapter 8) can be explained by negative costs agents perceive. Cost also arise from the situational structures of incentives in which the agents find themselves. Following other important values such as being successful in fulfilling one's tasks as a politician (for example reducing 'red zones' to enable landscape development) makes those strategies very costly that prevent exactly that task (as for instance afforestation). But it needs to be stressed once again that the understanding of costs in this approach is a comprehensive one. Given the same situational structure of incentives, another person with high ecological norms may feel it to be more costly not to follow such norms than to facilitate the minimisation of red zones on the map. Consequently, costs in this understanding not only prevent protection measures, they can also promote them.

For instance a person searching for adventure and ready to ski down a side run (motivation), who also feels able to take the steep slope, might imagine the disapproval such an action would earn from her friends. The cost of their social sanctions can outweigh the fun of skiing in deep snow and unprepared runs.

6 THE REAPPRAISAL PROCESSES

Appraisal of motivation and competence is a process that is repeated again and again and often modified during each reappraisal stage until the problem is either solved or finally felt to be unsolvable or irrelevant.

Assuming a situation where individual motivation and competence are in accordance with each other a clear intention can be formed; whether such intentions are actually realised is an other matter, since new, unforeseen barriers can always take effect. Such barriers are then evaluated in the subsequent reappraisal phase.

But if the outcome of the appraisal of competence, which encompasses the perception of response efficacy, self-efficacy and costs, is not in accordance with motivation, no intention to act is generated, and the re-appraisal starts at once. The outcome of the assessment of competence influences the appraisal of the motivation in the subsequent appraisal phase.

Sometimes situations may appear to be less threatening than at first perceived. For a person who perceives himself or herself to have a low ability to do something about a threat, a reinterpretation ('After all, things are not that bad ...') is a suitable way to decrease the motivation to do something as well. This is a typical strategy seen, for example, among people living in earthquake prone regions (Lehman and Taylor, 1987). Those persons who perceive a low competence due to high costs are usually in a better position than those who perceive low response or self-efficacy, since perception of costs is often more permeable: The high costs of doing something for the environment can quickly become an acceptable price to pay, if one's own life is at stake. For the enhancement of response or self-efficacy, external support is often necessary. Changes in the perception of a situation as 'very threatening' to 'a little threatening' can switch back to 'very threatening' again, if for instance new events in the environment occur. If the person still feels that no suitable protection measures are available, often an 'avoiding coping style' is the only possibility of dealing with the situation. Those strategies do not change the situation itself. They do, however, generate cognition and emotions that make the situation controllable and bearable for the persons involved. Examples are fatalistic thinking, religious faith and denial.

The perception of a situation can also change from threatening to challenging, if the person perceives herself or himself to be highly capable. The opposite is also true, of course.

In the process of such appraisal and reappraisal, new events, new information and of course the behaviour of other persons are considered and integrated. This change produces the movement from one type-box to the other as indicated in Figure 6.1.

7 OVERVIEW AND SUMMARY: THE PROCESS OF MOTIVATION AND COMPETENCE APPRAISAL

All the appraisal processes described above represent the human-nature interaction model and explain why humans are more or less successful at implementing prevention and protection measures. In summary, Figure 6.4 depicts the schematic sequence of the whole appraisal process: Depending on the specific social environment, on specific personality and on the characteristics of the environmental situation, persons perceive a given environment in different ways. However, there are limits in the differentiation of perceptions. The clearer the signs of danger are (for example the disaster is already in process), the more unlikely it is that somebody sees that situation as irrelevant. Apart from such extreme situations, a broad range of perceptions is likely. Some perceive the same situation as immediate and directly threatening, others do not. Those who feel immediately threatened will be highly motivated to look for coping mechanisms and will start this process directly. Those who perceive a situation as threatening in the long term or as challenging will possibly feel responsible, for prevention and protection measures as well. If they do not feel responsible, a new appraisal process starts whereby the situation may be perceived as irrelevant. Consequently, no environmental protective behaviour is shown. These appraisal components represent the motivation part of the human-nature interaction model.

In a further appraisal step, the agents then check whether they can cope with the problem. If no coping strategy appears possible, it is unlikely that action results. Often, such persons choose the above-mentioned 'avoiding coping strategies'. Based on the results of the first run, they start the next run of the appraisal process. Those persons who feel able to do something evaluate the costs of the actions. If the costs are perceived as too high compared to the intensity of motivation, they start the next appraisal phase. Those persons who feel that the costs are in accordance with the motivation generate an intention to put their strategies into practice. Possibly, further barriers impede the implementation, hence during the subsequent reappraisal phase, these new experiences are taken into account. Some persons can overcome all the difficulties and implement their planned actions. They then take the results of their actions into account in the reappraisal phase as well.

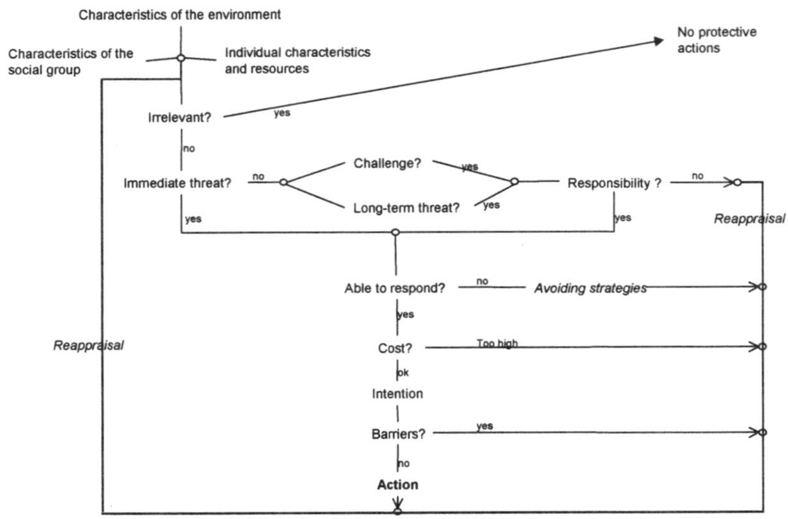

Source: Based on Amelang and Bartussek, 1985.

Figure 6.4: The human-nature interaction model: schematic sequence of the appraisal process of motivation and competence

8 CONCLUSIONS

It has been shown that many factors can impede the implementation of prevention and protection measures. These include different knowledge and value-based perceptions of the environment as well as institutional action entitlements and the perceived opportunities for actions. A diversity of perceived realities results and this can slow down the realisation of protection measures, particularly those that require a co-ordinated group decision, although clearly individual protection capacity is influenced by such perceptions as well.

The knowledge of such appraisal processes highlights several factors, which are relevant to deriving measures to promote protection capacity within a social group.

Empowerment of stakeholders

Assuming the case that there are different stakeholders' realities, this should be recognised and accepted. If the social group is interested in ensuring that not only the most powerful stakeholders follow their reality, measures need to be implemented that empower all relevant agents (for instance affected citizens, non-governmental organisations) in order to negotiate a common future reality. Many discursive approaches, like mediation or round tables, are known and have proven useful (see for example WBGU, 1999; Renn and Oppermann, 1995). However, the effectiveness of such participatory instruments presupposes the support of powerful agents (for instance a political board) or must be mandatory by law (for instance provided by the Convention on the Alps). There is a real danger that less powerful agents are already failing to acquire the means necessary for successful implementation of their wishes.

Convincing stakeholders

It has been shown that various factors (see Figures 6.2 and 6.3 for an overview) influence motivation and competence appraisal processes and whether subsequently intentions regarding protection and prevention measures are carried out or not. Those factors must be properly addressed if people are to be persuaded to generate protection capacity. The planning of strategies aimed at changing behaviour is well documented in environmental psychology research and would need an extra contribution (for overview: Homburg and Matthies, E. 1998; Mosler and Gutscher, 1998; Kotler and Roberto, 1991). Nevertheless, two basic rules are named here:

1. Strategies must be tailored to the requirements of specific target groups.
2. A diversity of strategies is necessary.

Measures to convince people can be more effective if they address the specific situation of the recipients. It is therefore necessary to analyse which persons or groups oppose effective protection measures and why. The human-nature interaction model presented above (see Figure 6.4) can be used to check which factors in any specific case inhibit effective protection measures: This raises questions such as the absence of a common understanding about how threatening a situation is or whether better knowledge can overcome this? Do the relevant persons reject specific protection measures because the measures endanger central interests and values? Can those interests and values then be changed directly, for example by a move to minimising life loss instead of having less red zones or indirectly, for example by enhancing public awareness and creating public pressure for protection measures? Can some form of compensation be found for those who do not follow their original interests?

It is likely that for each target group several factors act simultaneously to reduce the motivation and the competence to develop a high protection capacity (see Figures 6.2 and 6.3). All relevant factors such as other people's expectations (social norms), values, institutional entitlements, specific situational factors an so on must be addressed at the same time in order to promote a higher protection capacity effectively.

Reducing institutional barriers

Institutional designs like laws, voluntary agreements, control and sanction mechanisms act as external factors that may further or hinder the development of protection capacity. This point is of immense importance and demands particular attention. It has been included indirectly in the issue 'institutional entitlement' discussed above. Collective protection capacity is more powerful at protecting both the human social system and the ecosystem. Therefore the demands are put on persons who have institutional entitlements particularly high when it comes to altering the situational factors towards more sustainability. Questions of relevance here might be, how external control mechanism can ensure that existing risk management strategies are properly implemented at the local level, or how economic and ecological interests can be brought together by renewing subsidies systems.

In this contribution it has been shown that psychological factors such as social norms, values, knowledge, motivation and competence all influence the capacity to develop and implement protection measures. Only if – in addition to institutional and external barriers – such factors are taken into account can the vulnerability of the ecosystem and the human social system be reduced.

BIBLIOGRAPHY

Blaikie, P.T., Cannon, I.D. and Wisner, B. (1994). *At Risk: Natural Hazards, People's Vulnerability and Disasters*. London: Routledge.

Bloebaum, A., Hunecke, M., Matthies, E. and Hoeger, R. (1998). *Die Integration von oekologischer Normorientierung und Situativen Faktoren* (Forschungsbericht). Bochum: Fakultaet fuer Psychologie, Kognitions- und Umweltpsychologie.

Bohle, H.G., Downing, T.E. and Watts, M.J. (1994). Climate change and social vulnerability: toward a sociology and geography of food insecurity. *Global Environmental Change*, 4 (1), 37–48.

Bonß, W. (1995). *Vom Risiko: Unsicherheit und Ungewißheit in der Moderne*. Hamburg: Hamburger Edition.

Cutter, S. L. (1996). Vulnerability to environmental hazards. *Progress in Human Geography*, 20 (4), 529–539.

Edwards, W., and von Winterfeld, D. (1985). Public disputes about risky technologies: stakeholders and arenas. In V.T. Covello, J.L. Mumpower, P.J.M. Stallen, and V.R.R. Uppuluri (Eds.), *Environmental Impact Assessment, Technology Assessment, and Risk Analysis: Contributions from the Psychological and Decision Sciences* Berlin: Springer, pp.877–915.

Fuhrer, U. and Woelfing, S. (1997). *Von den sozialen Grundlagen des Umweltbewußtseins zum verantwortlichen Umwelthandeln – Die sozialpsychologische Dimension globaler Umweltproblematik*. Bern: Huber.

Gardner, G.T. and Stern, P.C. (1996). *Environmental Problems and Human Behavior*. Boston: Allyn and Bacon.

Guagnano, G.A., Stern, P.C. and Dietz, T. (1995). Influences on attitude-behavior relationships: a natural experiment with curbside recycling. *Environment and Behavior*, 27 (5), 699–718.

Homburg, A. and Matthies, E. (1998). *Umweltpsychologie – Umweltkrise, Gesellschaft und Individuum*. Weinheim und Muenchen: Juventa.

Jungermann, H. and Slovic, P. (1993). Die Psychologie der Kognition und Evaluation von Risiko. In Bechmann, G. (Eds.), *Risiko und Gesellschaft. Grundlagen und Ergebnisse interdisziplinaerer Risikoforschung*. Opladen: Westdeutscher Verlag, pp.167–207.

Kannapin, O., Pawlik, K. and Zinn, F. (1998). Praediktormuster selbstberichteten Umweltverhaltens. *Zeitschrift fuer experimentelle Psychologie*, 45 (4), 365–377.

Kelly, P.M., and Adger, W.N. (1999). *Assessing Vulnerability to Climate Change and Facilitating Adaptation*. Working Paper GEC 98-02, Centre for Social and Economic Research on the Global Environment, University of East Anglia and University of London.

Kroemker, D. and Mosler, H.-J. (2000). *Environment and Security: Coping Strategies in the Face of Environmental Scarcities*. Paper presented at the 3rd Biennial Conference of the European Society for Ecological Economics, Vienna 3–6. May.

Kotler, P. and Roberto, E. (1991). *Social Marketing*. Duesseldorf: Econ.

Lazarus, R.S., and Folkman, S. (1984). *Stress, Appraisal, and Coping*. New York: Springer.

Lehman, D.R., and Taylor, S.E. (1987). Date with an earthquake: coping with a probable, unpredictable disaster. *Personality and Social Psychology Bulletin*, 13 (4), 546–555.

Mayntz, Renate (1996), Das Expertendilemma. Politik und Wissenschaft – ein Spannungsverhaeltnis. In *Spektrum der Wissenschaft*, May 1996, 34–37.

Mosler, H.-J. and Gutscher, H. (1998). Umweltpsychologische Interventionen fuer die Praxis. *Umweltpsychologie*, 2 (2), 64–79.

Renn, O. (in press). Risikokommunikation: Umwelt und Gesundheit. In Ministerium fuer Umwelt und Verkehr Baden-Wuerttemberg (Eds.), *Kommunikation ueber Umweltrisiken. Zwischen Verharmlosung und Dramatisierung*. Tagungsband zum Symposium vom 30.11.2000 in Stuttgart. Stuttgart: Hirzel.

Renn, O. and Oppermann, B. (1995). 'Bottom-up' statt 'Top-down' – die Forderung nach Buergermitwirkung als (altes und neues) Mittel zur Loesung von Konflikten in der raeumlichen Planung. Stadtoekologie. *Zeitschrift fuer Angewandte Umweltforschung* (Sonderheft 6), 257–276.

Rippetoe, A.P. and Rogers, R.W. (1987). Effects of components of protection-motivation theory on adaptive and maladaptive coping with a health threat. *Journal of Personality and Social Psychology*, 52 (3), 596–604.

Schwartz, S.H. and Howard, J.A. (1981). A normative decision-making model of altruism. In J.P. Rushton and R.M. Sorrentino (Eds.), *Altruism and Helping Behaviour.* Hillsdale: Erlbaum, pp. 189–211.

Shrader-Frechette, K.S. (1995). Evaluating the expertise of experts. *Risk: Environment, Health and Safety,* 6 (1), 115–126.

Slovic, P., Fischhoff, B. and Lichtenstein, S. (1985). Rating the risks: The structure of expert and lay perceptions. In Covello, V.T., Mumpower, J.L., Stallen, P.J.M. and Uppuluri, V.R.R. (Eds.), *Environmental Impact Assessment, Technology Assessment, and Risk Analysis*. Band 4. Berlin: Springer, 131–156.

Tafarodi, R.W. and Walters, P. (1999). Individualism-collectivism, life events, and self-esteem: a test of two trade-offs. *European Journal of Social Psychology*, 29, 797–814.

Wissenschaftlicher Beirat der Bundesregierung Globale Umweltveraenderungen (WBGU) (1999). *Welt im Wandel: Strategien zur Bewaeltigung globaler Umweltrisiken – Jahresgutachten 1998.* Berlin: Springer.

7. Developments in Natural Hazard Management in Alpine Countries Facing Global Environmental Change

Johann Stoetter, Gertraud Meissl, Alexander Ploner and Thomas Soenser

1 INTRODUCTION

Due to the relief energy in geologically young mountain areas like the Alps, periodic and episodic mass movements, avalanches and flood events are common natural processes. It is the exposure of people and property, i.e. contact with the human spheres of interests, that make these natural processes hazards.

The world-wide trend of growing risks associated with natural hazards reported by reinsurance companies (e.g. Muenchener Rueckversicherung 1999) seems to be supported on a regional scale in Tyrol where several extreme events have occurred within the last two years: the debris flow onto the Brenner motorway ('*Brennermure*') in August 1998, the avalanche catastrophe in Galtuer and Valzur in February 1999 and the floods in May 1999. Other hazardous events were for example the rockfall from Eiblschrofen in June 1999 and the landslide in Schoenwies in August 1999.

An explanation for this frequent occurrence may be seen in the many and complex changes, both in the natural environment and the environment formed and controlled by human activities, that have been observed since the mid-20th century in particular. While growing population density and greater mobility lead to expanding settlements, infrastructure, industrial facilities and tourism, there may be an intensification of process dynamics due to changing climatic conditions.

The resulting rapid change in both the character of the natural hazards and the associated risks calls for a new approach in managing this situation.

2 DEVELOPMENT OF NATURAL HAZARD MANAGEMENT IN THE ALPS

Dealing with natural hazard processes has a long tradition in most countries in the Alps. The first official authorities for the protection against natural hazards were founded in the second half of the 19th century, e.g. in Switzerland in the late 1870s (Frutiger 1980) and in Austria in 1884 (Bergthaler 1975). For more than half a century technical defence works were developed and put in place. These 'active' measures, which represent the human *reaction* to hazard processes, appeared to be the appropriate way to cope with this challenge.

There was little impetus behind any new forward thinking prior to the 1950s and 1960s, when extreme avalanche and debris flow events occurred over wide areas of the Alps. Calculations of the financial requirements made it more and more obvious that there were financial limitations on these traditional reactive measures. This situation gave rise to the idea of complementary 'passive' protection measures, hence the idea of hazard zone maps was formulated.

Although the first avalanche zone map had been designed earlier in 1954 (for the village of Gadmen in the Berner Oberland), hazard zone maps were not explicitly mentioned in legal decrees before the mid-1960s (Baumann and Buri 1994). Both the *'Kantonales Forstgesetz'* (1963) of Graubuenden and the *'Vollziehungsverordnung zum Bundesgesetz betreffend die eidgenoessische Oberaufsicht ueber die Forstpolizei'* (1965), however, are restricted to avalanche hazard zones.

The Austrian forest law of 1975 represented a more sophisticated way of hazard zoning, which has been applied successfully over the whole country since then. The 'Verordnung ueber die Gefahrenzonenplaene' of 1976, which is related to the forest law, exactly defines the procedure for establishing hazard zone maps. Thus, for the first time, hazard zoning became standardised for a whole state.

A comparison of hazard zone concepts pursued in different countries in the Alps reveals that the longer the tradition the more the procedure is dominated by a single scientific discipline (Stoetter et al. 1999). In Austria the forest law of 1975 restricts all hazard zone planning to forestry engineers with a special education; in France, the experts responsible for these plans are predominantly geologists (Besson 1996 pers. com.).

Only in some of the more recent scientific approaches is the idea of an interdisciplinary approach for the management of natural hazards taken into consideration (Belitz et al. 1996; Ploner and Soenser 1997, 1999). However, the situation of official authorities in most countries shows that a monodisciplinary approach is still common, e.g. in the Autonomous Province of Bolzano-Southern Tyrol where Art. 66/2 of the *Raumordnungsgesetz* of 1998 defines geologists or holders of a Ph.D. in agriculture or forestry as the only authorised group to deal with natural hazards (Autonome Region Trentino-Suedtirol 1998).

Since the 1990s a number of further developments in dealing with natural hazards have been observed in the countries of the Alps. The following aspects may be highlighted:

- Development of information systems on natural hazards on a regional scale.
  In Tyrol (Schabl 2001, pers. com.), Kanton Berne (Bollinger et al. 2000) and the Autonomouos Province of Bolzano-South Tyrol (Fuchs et al. 2001) concepts for such a tool for planning at a regional scale have recently been developed, and wider application initiated.
- Development of the concept of risk analysis.
  In contrast to the traditional way of assessing only the hazard process, a new risk-based hazard zoning concept (see Heinimann et al. 1998) considers the value and probability of presence of goods and persons. This reveals totally new aspects for natural hazard management in alpine environments.

3 GLOBAL CHANGE – CHALLENGE OF NEW DIMENSION

Global change is frequently equated with climate change. The authors of this chapter, however, would like it to be understood as a term for all changes, both in the natural and the humanmade environment, that have occurred since the beginning of industrialisation and that will occur in the future. Technical innovations, improved medical care and an enormous increase of mobility have provided the stimuli for a rapid change in society as a whole. As a consequence of an increasing population, all spheres of human interests have grown in importance and in terms of demand for space.

In addition to this, it is the traceable human impact on the climate system through the additional emission of greenhouse gases and through intensified humanmade/technical environmental degradation that marks the beginning of global change on the natural environment side.

Changes in the Natural Environment

In a pure natural environment all changes are either forced by climate change processes or related to natural successions. Due to the many and complex human activities it has become almost impossible to identify distinct natural environments apart from those under human impact, e.g. by changing the landuse. Thus, most of the vegetated areas in the Alps, although they might appear to be natural environments, are in fact humanmade environments.

In this chapter only changes in the natural environment forced by climate change processes are discussed. Although the predicted magnitude of future climate change has varied throughout the last decade there is strong evidence for major positive trends of both temperature and precipitation (Houghton et al. 2001). Due to a better understanding of the complex processes of energy exchange between atmosphere, water and land, and to the availability of more sophisticated climate models, scenarios of high probability and reliability on a global scale have been produced.

There are still a lot of uncertainties at the regional scale; especially for mountain areas like the Alps a higher spatial resolution is necessary (Ohmura and Beniston 1996; Wanner et al. 2000). Nevertheless, all model results of scenarios of future climate conditions in Europe suggest a warming throughout the 21st century that is accompanied by a change in the precipitation pattern.

As some of the hazard processes are strongly related to precipitation, a change towards single precipitation events of higher intensity will produce e.g. debris flow processes with a higher energy potential.

A warmer atmosphere also has a direct impact on the cryosphere in the Alps. While snow reacts immediately to warmer conditions, both glaciers and permafrost bodies show reactions only after a time lag of years to decades (Stoetter 1994). These relationships can clearly be shown for the period since the Little Ice Age maximum in the mid-19th century. While over the Alps as a whole the glaciers have lost more than a third of their surface area during this period (Maisch et al. 1998), estimates of the spatial reduction of permafrost areas vary from 10% (Haeberli et al. 1998) to 40% (Stoetter et. al. 1996). A further warming of 1K will cause a spatial reduction of both glaciers and permafrost comparable to that observed within the last ca. 150 years (Maisch et al. 1998; Haeberli et al. 1998). The main effects of this loss of ice on hazardous processes are that (a) large areas of unstable, loose material will be freshly exposed to erosion processes and (b) the release zones of the resulting mass movement processes will be at a higher altitude causing higher energy potential and process dynamics.

Table 7.1: Process dynamics of natural hazard processes under changing climatic conditions

Process	Probability of occurrence	Magnitude of impact	Impact on
Occurrence of avalanches – change	–	(±)	People, buildings, goods, infrastructure
Debris flows (high Alps)	+	–	Buildings, goods, infrastructure,
Debris flows (Alpine margin)	±	–	environment
Runoff regime with more winter runoff	++	–	Environment, water resources, hydroelectric power
High water level	+	– – –	People, buildings,
Floods	++	– – –	goods, infrastructure
Summer storms/hail	++	–	Buildings, goods, infrastructure,
Winter storms	+	– –	agriculture, forestry

Notes:

Probability of occurrence due to climate change:

++ intermediate probability of occurrence;

\+ low probability of occurrence;

± no or very uncertain probability of occurrence;

– decreasing probability of occurrence.

Magnitude of potential impact (damage or benefit potential):

– – – considerable impact with negative consequences: high damage potential;

– – intermediate impact with negative consequences: intermediate damage potential;

– small impact with small damage potential;

\+ small impact with positive consequences: benefit potential;

() very uncertain impact.

Source: Modified from Bader and Kunz (1998), p.236.

There is however a positive impact on vegetation. The rise of the tree-line and vegetation boundaries may lead to restabilisation and protection against specific erosion processes. This development will, though, take decades to centuries. The more rapid reaction of the cryosphere opens a period of more unstable conditions before the system will become stable again.

Bader and Kunz (1998) demonstrate that with the exception of avalanches, all hazard processes mentioned show an increase in process dynamics (see Table 7.1).

Changes in the Humanmade Environment

Although human activity can have an impact on natural hazard processes at all levels, the authors concentrate on highlighting those human activities that increase the damage potential in areas affected by natural hazard processes.

The number of inhabitants in the Eastern Alps grew in almost all communities during the 20th century (Baetzing 1993). As this development was accompanied by a considerable increase in individual demand for space the settlement structure has changed markedly. Small villages, often lying in sheltered positions, have been turned into accumulations of buildings greater in number and bigger in size covering a much larger area than they did before.

The sociological aspect of a change from a society predominantly determined by agriculture to a society of mobility and leisure is of considerable importance for the development of the overall situation. This process is linked with:

(a) the demand for new infrastructure both in the old settlement centres (hotels etc.) and in areas previously avoided (ski lifts, car parks etc.);
(b) an additional high presence of people (tourists) at specific periods (holidays seasons, weekends), which multiplies the original Alpine population; and
(c) an almost complete loss of the awareness of natural hazards and the associated risks and, connected to that, a decrease in risk acceptance.

All these aspects of change in the humanmade environment lead to both a higher probability of the presence of human beings and property and thus to an increase in the value of items becoming exposed to hazard processes.

		Value/probability of presence (goods/persons)		
		decreasing	constant	increasing
Process-dynamics	increasing	∅	−	− −
	constant	+	∅	−
	decreasing	++	+	∅

Figure 7.1: Possibilities of future risk development due to changes in the natural and humanmade environment.

As pointed out above (see e.g. Table 7.1), there will be very few positive impacts of climate change on the hazard processes in the natural environment in the immediate future. It is also unlikely that the economic values or the number of people exposed to hazard processes will decrease. Instead the demand for both recreation and mobility will increase, thus causing a higher probability of presence. Therefore, the overall perspective for the future development of the risk situation is not very optimistic; there is a high probability of a concentration in the upper right part of the matrix in Figure 7.1, i.e. of increasing risks.

Since the initial founding of official authorities for the protection against natural hazards, extensive defence works have been established. It is estimated that the total value of these constructive measures in Austria is about EURO 10 billion (Weinmeister 2000). Apart from the demand to assess the functionality of some of these technical constructions under the light of the present state of the art, the simple assumption of an average lifetime of 100 years for such constructions means additional costs of EURO 100 million per year just to keep the security level at its present state. This sum does not however include any financial investment to provide new measures for the protection from present hazards, let alone those expected for the future due to the changes in nature and society discussed above. Against the background of more and more limited public financial resources, this means that an overall increase of the effectiveness of the defence works is not really plausible. Thus a positive future development of the risk situation (lower left part of Figure 7.2) is more or less unlikely.

		Effectiveness of defence works		
		increasing	constant	decreasing
Global change scenario risk	strongly increasing	–	– –	– –
	increasing	∅	–	– –
	constant	+	∅	–
	decreasing	++	+	∅
	strongly decreasing	+++	++	+

Figure 7.2: Possibilities of future risk development due to complex global change processes according to the five categories in Figure 7.1 and the future effectiveness of the defence works

4 SUSTAINABLE PROTECTION OF THE ALPINE *LEBENSRAUM*

In view of the challenge of rapidly changing basic conditions both in the natural and the humanmade environmental system, sustainable protection of the Alpine *lebensraum* must be the main goal for all responsible authorities in the Alps. This can be achieved by meeting the following demands.

Demand 1

As natural hazards are complex phenomena affecting many aspects of both the natural and the humanmade environment, sustainable solutions require an interdisciplinary and cross-disciplinary approach.

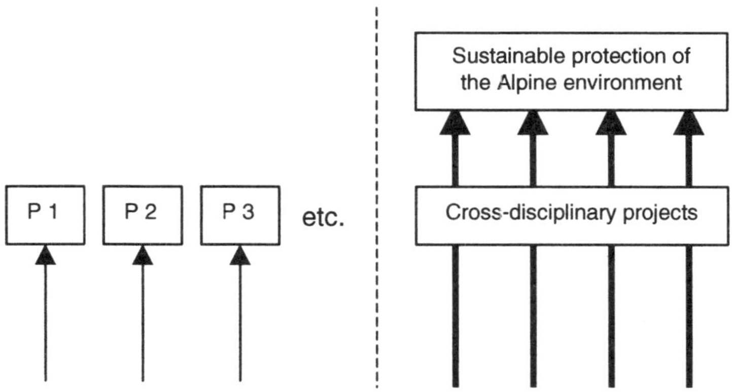

Figure 7.3: Mono-disciplinary projects (P1, P2, P3 etc.) versus cross-disciplinary approach

This requirement is in stark contrast to the present situation, where in most of the countries of the Alps the official authorities dealing with natural hazards are controlled by graduates of a specific discipline, e.g. engineers, forestry engineers or geologists. Although all these officials are trained specifically in natural hazards, the gap between the complex problems natural hazards now present and the monodisciplinary training background cannot be closed. It is quite natural that:

(a) single disciplines specialise in very specific aspects of the overall topic;
(b) research activities and further developments are conducted by one

discipline without considering advances by other disciplines (see Figure 7.3).

Demand 2

All groups affected must be involved, either by active participation or by appropriate information management.

Due to their legal status as the exclusive bodies dealing with natural hazards, the official authorities have increasingly evolved into administrative bodies. Natural hazards, however, are not a matter of administration but are essential problems that affect the fundamental living conditions of the entire society.

Demand 3

Active foresighted solutions to prevent hazards or their effects must be developed taking into account the changing basic conditions both in the natural and the humanmade environment.

As shown above, the way of dealing with natural hazards in the Alps to date has been mainly reactive. The events mentioned in the introduction are typical examples from the recent past. When changed basic conditions both in the natural and the humanmade environment, not previously considered, caused unexpected events, protective measures were immediately established. Reactive measures have dominated while active planning with regard to the future has mostly been missing. Thus the overall demand for living with natural hazards is to develop and strengthen a new approach towards natural hazard management.

5 IDEAS FOR A FUTURE NATURAL HAZARD MANAGEMENT IN TYROL

There are different ways to address this problematic situation. Mere declarations will not be of any help, however, even if they focus directly on the issue like the Vienna Declaration at the CALAR conference (initiated by the EC-approved project Concerted Action on Forecasting, Prevention, and Reduction on Landslide and Avalanche Risks) of January 2000, which lists items for specific action in order to reach the two overall goals that are:

(a) better preparation of society for the impact of disasters;
(b) proceeding from reaction to and protection against hazards, to the

management of risk by integrating risk prevention strategies into sustainable development programmes.

Instead of oral statements, tangible activities that work on the problem are required. As an example, an initiative in Tyrol is presented in the following section.

In Tyrol a committee of interested parties from the University of Innsbruck (co-operating with other university institutes and scientific institutions), governmental offices of both the state of Austria (Torrent and Avalanche Control and Federal Forest Research Centre) and the region of Tyrol (Government of Tyrol) and from private companies, has suggested the establishment of a centre of intelligence for the development of a new approach to natural hazard management in order to create an essential base for facilitating sustainable protection of the Alpine *lebensraum*.

On 20th August, 2001 a proposal for the installation of a centre of intelligence named 'alpS – Centre of Natural Hazard Management' (Kompetenzzentrum 'alpS – Zentrum fuer Naturgefahren-Management') was submitted to the Austrian Federal Ministry for Traffic, Innovation and Technology (Bundesministerium fuer Verkehr, Innovation und Technologie). Following a review process, a decision about the implementation of the centre within the framework of the Kplus-programme, which aims at improving co-operation between scientific institutions and private companies, will be made later in 2002.

It is intended that research in the planned centre of intelligence complies with the demands stated above and that it improves:

(a) co-ordination of ongoing activities in administration, research and the economy;
(b) exploitation of existing knowledge;
(c) the creation of new knowledge/new expertise.

It has been proposed that the 'alpS – Centre of Natural Hazard Management' takes over a leading role in the field of natural hazard management and provides an impulse for the development of new products.

In order to fulfil this task the internal structure of the planned centre of intelligence will be designed to represent the different disciplines involved in research on natural hazard management. The following three areas of research have been defined (see Figure 7.4).

Research Area A: Database and Modelling

The establishment of databases on natural hazard data needs to consider all aspects from the acquisition, administration and processing of the data to their interpretation and reporting. Comprehensible, comparable and reproducible decision-making processes require access to standardised, high-quality data at different scales. Only data of a sufficient spatial and temporal resolution allows event- and site-specific assessments, demarcations and the recommendation of suitably proportioned hazard mitigation and protection measures. The further development of models aims both at a better understanding and representation of the physical processes, and at improving the practical application of the modelled processes by means of comprehensible algorithms and representative parameters.

As the present situation related to data availability and model suitability is, however, far from meeting these demands and objectives, area A is designed to meet the following objectives: On the basis of a systematic compilation and evaluation of the databases, methods of data acquisition and models available, further developments will be aimed at increasing their effectiveness and efficiency in order to provide standardised high-quality databases and modelling results. As far as hazard assessment is concerned, a paradigm change will be implemented in the definition of event prediction in order to supplement the traditional analysis of statistical records, which are often too short, by system- and process-oriented approaches. A further task within area A will be to develop scenarios of natural hazard situations under changed frame conditions.

Research Area B: Hazard Mitigation and Protection Measures

Research in area B deals with the planning strategies and implementation of structural and non-structural protection measures against hazard events like torrential floods, debris flows, landslides, avalanches and rockfalls. Against the background of a constantly changing natural and humanmade environment, i.e. a changing risk situation, the type and implementation of these measures must be adapted to varying scenarios. In order to improve the approach now adopted to cope with natural hazards, which is predominantly reactive, area B is designed to meet the following objectives.

Based on a systematic compilation and evaluation of the protective measures used at present, process- and risk-oriented approaches for constructive and bioengineering, planning and temporary measures will be developed. In this development, a new orientation away from the traditional reactive form and towards more preventive measures will necessitate the implementation of a paradigm change in the planning of measures. In particular, preventive measures will be developed for natural hazard situations under global change conditions.

Research Area C: Socioeconomic Risk Analysis

Research in area C focuses on socioeconomic aspects of risk analysis, i.e. the overall process of risk perception, risk assessment, risk management and risk communication (including disaster management). The emphasis is placed on the economic assessment of natural hazards and protective measures in mountain regions (especially the Alps), on the insurability of objects and persons against natural hazards and on psychological aspects of risk perception, acceptance and communication, as well as on coping with the damages.

Based on the compilation and evaluation of the socioeconomic models that have been used in natural hazard management, new developments will focus particularly on socioeconomic assessment of potential damages as well as on developing strategies for risk communication, disaster prevention and management. Development of adequate strategies for decision making related to natural hazards will support the paradigm change implemented in area B. For global change scenarios, socioeconomic assessment of potential damages and risk management strategies will be suitably elaborated upon.

As 'alpS – Centre of Natural Hazard Management' is intended be a place for interdisciplinary co-operation, strong links between the research fields are vital, and research projects will usually cover more than one research field. By ensuring interdisciplinary research as well as requiring co-operation between science and practical work by governmental offices and private companies, the centre of intelligence can establish new, more effective and more efficient methods of dealing with natural hazards.

The results from the research and development programme of the 'alpS – Centre of Natural Hazard Management' are expected to:

- increase the efficiency and effectiveness of natural hazard management;
- necessitate implementation of a paradigm change;
- develop strategies for natural hazard management under changed basic conditions in the future.

The overall goal of sustainable natural hazard management can only be achieved, however, if up-to-date information about the research activities and the latest results from the centre are disseminated to as wide a public as possible, as can be shown by two examples:

- Experts in charge of the planning of measures need to know about newly available data, models, methods of planning etc. to be able to utilise and apply them.
- Disseminators like teachers or ski instructors need to have access to newly developed media for risk communication so as to be able to improve the risk perception of children and young people, and to lead group discussions about risk behaviour and risk acceptance suitable for a range of age - groups.

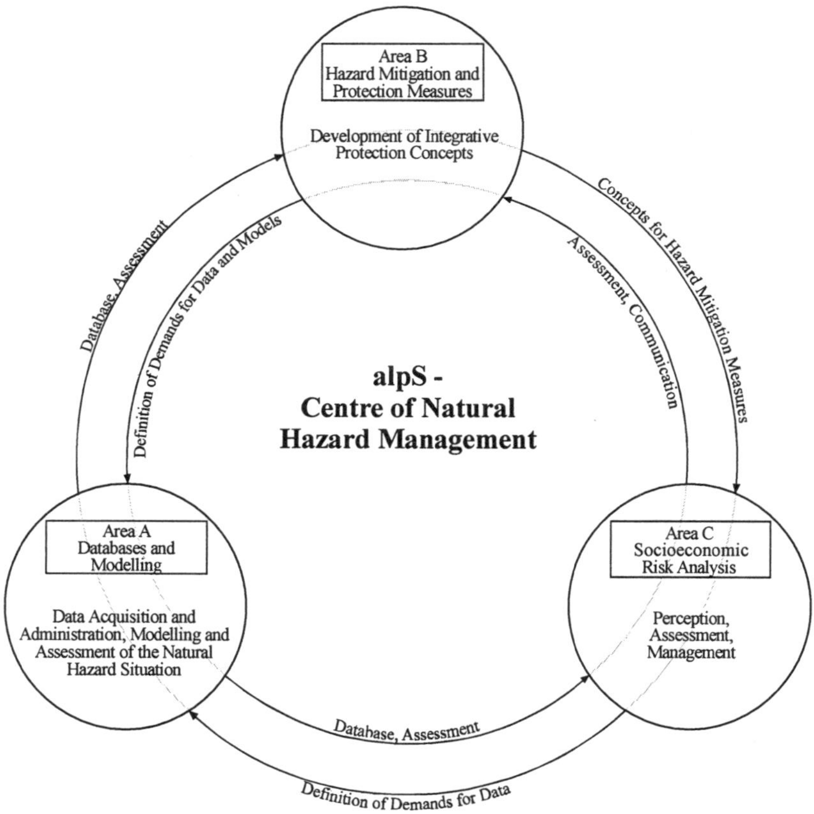

Figure 7.4: Structure of the planned centre of intelligence and interrelation between the research areas

To facilitate the above, the activities of the 'alpS – Centre of Natural Hazard Management' will be accompanied by the launch of an educational programme to communicate successfully an adapted and sustainable approach to natural hazard management, comprising scientific, technical, economic, legal and psychological aspects of relevance to the different target - groups.

Tyrol is a natural site for a centre of intelligence dealing with natural hazards as the entire country lies in the Alps and is confronted with all kinds of alpine natural hazards. Due to this fact a wide range of institutions working on natural hazards is already situated in Tyrol, thus making it an ideal basis for a centre of intelligence.

Whether the centre of intelligence will come into being in Tyrol, either within the framework of the Kplus-Programme of the Austrian Federal Ministry for Traffic, Innovation and Technology or through another source of funding, also depends on political and economic decisions. Irrespective of the result of these decisions, behind the initiative of founding such a centre there lies a firm idea to improve natural hazard management by meeting the demands stated above.

6 CONCLUSION

Given the background of changing basic conditions in the natural and humanmade environment outlined above, it is clearly now time to act rather than discuss. The establishment of centres of intelligence dealing with natural hazards is strongly dependent on the willingness of the political decision-makers. In view of the obvious facts that (a) global change has an impact on natural hazards and associated risks and (b) the foreseeable failure of the traditional concepts as used by official authorities, it seems logical that political decision-makers have to favour such a concept as it is one of the fundamental responsibilities of elected representatives to take care of the welfare of society.

ACKNOWLEDGEMENTS

The authors thank Dr. Maria Wastl and Dr. Chris Caseldine for revising the English, Mag. Margreth Keiler for drawing Figure 7.4, and both the editors and two unknown reviewers for valuable comments.

BIBLIOGRAPHY

Autonome Region Trentino-Suedtirol (1998), 'Beiblatt Nummer 5 zum Amtsblatt der Autonomen Region Trentino-Suedtirol vom 28. April 1998', **18-I/II**, Trient.

Bader, S. and P. Kunz (1998), *Klimarisiken – Herausforderung fuer die Schweiz*, Wissenschaftlicher Schlussbericht NFP31, vdf Zuerich.

Baetzing, W. (1993), *Der sozio-oekonomische Strukturwandel des Alpenraumes im 20. Jahrhundert. Eine Analyse von 'Entwicklungstypen' auf Gemeinde-Ebene im Kontext der europaeischen Tertiarisierung*, Geographica Bernensia, **P 26**, Bern.

Baumann, R. and H. Buri (1994), 'Erfahrungen mit den Richtlinien zur Beruecksichtigung der Lawinengefahr', *Informationshefte Raumplanung*, **1**, 29–30.

Belitz, K., U. Frisch, M. Maukisch, J. Stoetter and O. Baume (1996), 'Entwicklung eines Konzepts fuer einen Gefahrenzonenplan fuer Gemeinden des bayerischen Alpenraums – Recherche zu Konzepten im oesterreichischen, schweizerischen, franzoesischen, italienischen und bayerischen Alpenraum und Konzeptvorschlag', *Studie im Auftrag des Bayerischen Landesamts fuer Wasserwirtschaft*, Muenchen, Institut fuer Geographie der Ludwig-Maximilians Universitaet Muenchen.

Bergthaler, J. (1975), 'Grundsaetze bei der Erarbeitung von Gefahrenzonenplaenen in Wildbaechen der Noerdlichen Kalkalpen und der Grauwackenzonen', *Oesterreichische Wasserwirtschaft*, **7/8** (27), 160–168.

Bollinger, D., H. Buri, G. della Valle, C. Hegg, H.R. Keusen, H. Kienholz, B. Krummenacher, P. Mani and H. Roth (2000), Gefahrenhinweiskarte des Kantons Bern 1:25000, *Tagungsband INTERPRAEVENT* (2000), **2**, 189–200.

CALAR Conference (2000), 'Vienna Declaration', http://www.swedgeo.se/calar/vienna_declaration_homepage.html

Frutiger, H. (1980), 'Schweizerische Lawinengefahrenkarten', *Tagungsband INTERPRAEVENT*, **3**, 135–143.

Fuchs, S., M. Keiler and A. Zischg (2001), *Risikoanalyse Oberes Suldental, Vinschgau – Konzepte und Methoden zur Erstellung eines Naturgefahrenhinweis-Inforamtionssystems*, Innsbrucker Geographische Studien, **31**.

Haeberli, W., A. Kaeaeb, M. Hoelzle, H. Boesch, M. Funk, D. von der Muehll and F. Keller (1998), *Eisschwund und Naturkatastrophen im Hochgebirge*, Schlussbericht NFP 31, vdf Zuerich.

Heinimann, H., K. Hollenstein, H. Kienholz, B. Krummenacher and P.

Mani (1998), *Methoden zur Analyse und Bewertung von Naturgefahren*, BUWAL (ed.), Umweltmaterialien, **85**.

Houghton, J., Y. Ding, D.J. Griggs, M. Noguer, P.J. van der Linden and D. Xiaosu (eds.) (2001), *Climate Change 2001: The Scientific Basis*, Assessment Report of the Intergovernmental Panel on Climate Change (IPCC), Cambridge University Press, Cambridge.

Maisch, M., A. Wipf, B. Denneler, J. Battaglia and C. Benz (1998), *Auswirkungen von Klimaaenderungen auf die Gletscher und deren Vorfelder*, Schlussbericht NFP 31, vdf Zuerich.

Muenchener Rueckversicherung (1999), *Naturkatastrophen in Deutschland, Schadenerfahrungen und Schadenpotentiale*, Muenchener Rueckversicherungsgesellschaft, Muenchen.

Ohmura, A. and M. Beniston (1996), *Simulation of Climate Trends over the Alpine Region: Development of a Physically-based Modeling System for Application to Regional Studies of Current and Future Climates*, Final Report NRP 31, vdf Zuerich.

Ploner, A. and Th. Soenser (1997), Dynamische Gefahrenzonenplanung – ein Teil des Naturgefahren-Managements, dargestellt an Beispielen, *Tagungsband 'Risikobewertung und Naturraumpraevention von Wildbaechen und Lawineneinzugsgebieten,* Universitaet fuer Bodenkultur, Wien, 48.

Ploner, A. & Th. Soenser (1999), Natural Hazard zoning as an important basement of the evaluation of forest protection functions in Austria. *Tagungsband IDNDR 'Risques naturels en montagne', Grenoble,* 431–436.

Stoetter, J. (1994), *Veraenderungen der Kryosphaere in Vergangenheit und Zukunft sowie Folgeerscheinungen. Untersuchungen in ausgewaehlten Hochgebirgsraeumen im Vinschgau (Suedtirol)*, Habilitationsschrift an der Fakultaet fuer Geowissenschaften der Ludwig-Maximilians-Universitaet Muenchen.

Stoetter, J., M. Maukisch, M., J. Simstich and K. Belitz (1996), Auswirkungen des zeitlich-raeumlichen Wandels der Permafrostverteilung im Suldental (Ortlergebiet) auf das Gefaehrdungspotential durch Erosionsprozesse. Tagungsband INTERPRAEVENT, **1**, 447–457.

Stoetter, J., K. Belitz, U. Frisch, T. Geist, M. Maier and M. Maukisch (1999), 'Konzeptvorschlag zum Umgang mit Naturgefahren in der Gefahrenzonenplanung – Herausforderung an Praxis und Wissenschaft zur interdisziplinaeren Zusammenarbeit', *Innsbrucker Geographische Gesellschaft Jahresbericht*, **1997/1998** (14), 30–59.

Wanner, H., D. Gyalistras, J. Luterbacher, R. Rickli, E. Salvisberg, and C. Schmutz (2000), *Klimawandel im Schweizer Alpenraum*, Final

Report NRP 31, vdf Zuerich.

Weinmeister, W. (2000), Contribution towards sustainable protection against natural hazards, *Tagungsband INTERPRAEVENT*, **1**, 85–96.

8. The Political Practice of Natural Hazards Control in Austria and the Question of Climate Change

Gerhard Weiss

1 INTRODUCTION: NATURAL HAZARDS CONTROL AND CLIMATE CHANGE

Within the current discourse on climate change, the impact of changing weather conditions is often discussed. The amount of precipitation may increase and, in particular, more extreme weather is expected, including storms, thunderstorms, droughts and heavy rainfall. This implies a growing threat from natural hazards like landslides, mud-flows, floods and avalanches in the Alps and elsewhere. How should those responsible for risk regulation respond to these increased uncertainties?

This chapter presents some thoughts on this question derived largely from an empirical analysis of the current political practice of torrent and avalanche prevention in Austria. In a five-year research project, the work of the torrent and avalanche control service (TACS; *Forsttechnischer Dienst fuer Wildbach- und Lawinenverbauung*) was studied in order to evaluate the policy instruments for natural hazards prevention, as applied in Austria. The study covered mountain forest restoration (*Schutzwaldsanierung*), hazard zone mapping (*Gefahrenzonenplanung*), and subsidy policy for torrent and avalanche control measures (*Wildbach- und Lawinenverbauung*).[1]

2 ADMINISTRATING RISK: THE ROLE OF BUREAUCRACY IN RISK REGULATION

Politics and Bureaucracy

Public agencies are important actors in the formulation and the implementation of concepts, norms and procedures for risk regulation.

As the administration elaborates proposals for laws, not only their knowledge, but also their value system and interests are incorporated into the regulation process (Peters, 1984). Conflicts of interest that have not been solved in the process of formulation often lead to vagueness, ambiguity and contradiction in the final law. If the underlying conflicts cannot be resolved in implementation, implementation deficits are the result (non-decision problems, Wolf, 1986: 145, 151).

In the implementation process, the administration strongly influences the interpretation of the laws, but they act within a policy network in which other actors are involved as well. Through this network, values and interests of parties affected by the regulations become effective. The outcome of the implementation process depends on the power of the actors involved (Scharpf, 1997). As a result, risk concepts and risk assessments of experts and official agencies are biased (Shrader-Frechette, 1995). Risks are conceptualised within the value system of the professional community (Dietz and Rycroft, 1987: 77).

Public bureaucracies not only follow formal goals (contents of laws, general principles of bureaucracy like legality, economy and expediency) but also informal goals, for instance: public agencies strive at maximising their budget or safeguarding their autonomy, field of competence and legitimacy; in order to minimise conflicts they regulate a sector rather in correspondence with than against the interests and values of their main clientele (selective clientelism); policy measures are the result of negotiations between the administration and the addressee (bargaining process) (e.g. Downs, 1967; Mayntz, 1985; Scharpf, 1997). Depending on the power of the groups concerned, the result of this negotiation process may be that the public agency serves the interests of the private group rather than controlling them (capturing, Wolf 1986; state failure, Jaenicke, 1986). As in industry, clientelistic behaviour is also found in the forestry sector (Glueck, 1976; Krott, 1990; Pregernig and Weiss, 1998). A change of structures and processes are unlikely because of the resistance of established actors who benefit from the existing situation.

If the problem-solving capacity is restricted, small problems are solved whereas bigger problems are not (catastrophe paradox, Prittwitz, 1990). In the work of a bureaucracy even a reversal of the rational principle of choosing means according to certain goals might occur: The administration may have certain instruments and procedures at hand and, with the informal goal of keeping their territory and legitimacy, they seek goals which help to legitimise their work (Luhmann, 1966; Downs, 1967; Edelmann, 1976). In their problem-solving efforts they apply their own professional values and routines and change policies only under high pressure (Sabatier, 1993).

Institutions and Procedures in Risk Regulation

Not only 'risk decisions', but all decisions are risky because decisions are always taken under uncertainty. For the administration, however, decisions have to be certain. Various procedures serve to transform the uncertainty inherent in the problem into decision-making certainty (Luhmann, 1993: 144f.). Setting standards, carrying out measurements and adopting mathematical procedures are used to rationalise the highly uncertain and value-laden decision on 'what is acceptable risk' (Wolf, 1986). Formal procedures make the decision apparently value-free.

Risk management tries to find a rational approach to safety decisions by structuring the problem and making the decision transparent. However, these rationalist risk models, which are developed by technicians, are often not accepted by the public. Laypersons' perceptions are often considered 'irrational' by technical experts (compare Starr, 1969; Starr, 1980: 4) but their behaviour can also be called 'rational in a different sense' (Perrow, 1984). We learn that the conceptions of risk of both laypersons and technical experts depend on their interests and values. Neither laypersons nor experts are immune to biases in risk perception (Fischhoff et al., 1981). In many risk debates the technocratic procedural approach is questioned (Nelkin and Pollack, 1980: 243–246). These existing procedures put value decisions into the hands of technicians who – in their daily work – are not aware of the political implications.

Risk decisions always involve values and a variety of group interests. As risk decisions include (potential) costs and benefits they strongly depend on which groups will have to bear the costs and who profits. The traditional bureaucratic approach applied by public agencies has deficiencies in terms of democratic principles because the interests of different groups are not considered evenly in the implementation process. In spite of their formal rationality the concepts applied are often biased towards powerful interests (Mazur, 1985). The analysis of power is crucial in risk debates and risk regulations, respectively (Compare, 1995).

The consideration of interests in risk regulation is a question of political procedures. The bureaucratic decision and the use of risk models is only one possible procedure. Alternatively, more inclusive decision processes have been proposed (e.g. Dienel, 1997; from a critical viewpoint, van den Daele, 1994; for natural hazards management in

Austria Rappold, 2001). Not least does a successful regulation depend on the trust of the public in the capability of public agencies (Conrad, 1989; Wynne, 1992).

Also in the assessment of natural hazards, procedures exist that are ostensibly 'objective'. Yet, the underlying normative assumptions of what risk is regarded as acceptable is hidden behind expertise (Weiss, 1999). Concerning the use of forests to protect against natural hazards, for instance, it appears that forest officials perceive risk distinctly differently from representatives of the torrent and avalanche control service in Austria. While the former primarily look at risks for the forest, the latter look at risks for communities and infrastructure. Consequently, the two institutions develop different measures and strategies to reduce the risk (Weiss, 2001). Beginning with the same knowledge, foresters from nature conservation groups and foresters from the public administration arrive at different answers to the question of whether mountain forests produce risks if left unmanaged. While conservationists would rather trust in the capability of natural ecosystems to regulate themselves, official foresters prefer to presume that ecosystems have to be managed to be stable (Weiss, 2000a: 360f.).

The concepts of risk analysis and risk management have been applied in the field of natural hazards (Kienholz, 1993; Hollenstein, 1995). These authors recognise the problem of deciding on 'What is acceptable risk?', but up to now conceptual approaches for coping with these problems have been lacking in natural hazards regulation. Alternative procedures – e.g. public participation – have not yet been studied thoroughly.[2]

3 CASE STUDY: THE TORRENT AND AVALANCHE CONTROL SERVICE IN AUSTRIA

The 'Forest-technical System' for Torrents and Avalanche Control

The forest-technical service for torrent and avalanche control was founded in the late 19th century when the Act for Torrent Control came into force after severe floods had occurred in large parts of the Austrian Empire in 1882. Since that time the service has been assigned to forestry jurisdiction and implements what is known as the forest-technical system of torrent and avalanche control. This aims at combining biological and construction measures in order to prevent natural hazards. Not only technical means should be applied to prevent damage; the land should also be used in such a way that hazards are avoided as well (precautionary principle). For instance, alpine pastures should not be overgrazed so as to avoid erosion or landslides; alpine meadows should be afforested if this can prevent avalanches; in forest management, clearcuts that might cause avalanches or rock-fall should be avoided. In the 1970s hazard zone planning, another precautionary instrument, was introduced by the TACS. By hazard zone planning, land use should not only be controlled in the catchment areas of torrents and avalanches but also in the hazard-prone areas themselves. In times of prosperity many new buildings were constructed in such areas, thus causing additional needs for prevention measures against hazards. In connection with land-use planning, the development of hazard-prone areas is to be prevented.

Conflicts in Land-use Regulation

The TACS has different instruments at hand: technical structures, biological measures, and land-use regulation. In practice the erection of technical structures takes preference over a change or restriction of land uses (afforestation of pastures; hazard zoning) (Weiss, 1999).

One important reason for this policy outcome is that the restriction of land use causes more conflicts with local users (e.g. farmers, developers). Furthermore, technical structures like check-dams have immediate effects, thus having higher political value. Politicians prefer immediate effects and are not only interested in physical impacts (less future damage) but also economic (increased property values) and symbolic benefits (proof that they 'care' about the safety of their municipality). The symbolic value of construction is higher because the protective effects are more obvious. Although afforestation would provide long-term enhancement of the communal safety, it is not adequately promoted, and while the TACS considers land-use regulation a central part of watershed amelioration, such measures are not strongly required.

The reasons for this lie in the distribution of power between the actors involved in the policy network. This can be demonstrated by the following prototypical example (Table 8.1): In projecting avalanche protection, typically the ideal strategy is to build technical constructions and plant a forest on suitable sites. As a rule, there are strong interests in realising technical constructions but afforestation is partly opposed. Not only the village/town (applicant) supports constructions, but also the county authority which informally represents influential regional interests. The constructions themselves are opposed only by those concerned about nature protection. Furthermore, rather powerful actors oppose afforestation of alpine pastures in such a project (farmers, hunters, agrarian officials and tourist managers) while only a few actors support such a land-use change (representatives of the ministry, the local forest office and the nature protection office).

Table 8.1: Typical attitudes of political actors towards avalanche control projects with and without afforestations

	Acceptance **	
Actors *	Technical constructions only	Afforestations in addition to constructions
Torrent and Avalanche Control Service local office	+	?
Applicant (e.g. commune, transportation agency)	+	0
Land owner (farmer)	0	–
Hunting rights holder	0	–
Tourism	+	–
Regional authority (county)	+	0
Agrarian official expert	0	–
Ministry of Agriculture (Dept. of TACS)	+	+
Forestry official expert	0	+
Nature and landscape conservation official expert	–	+

Notes:

* Interest groups are ranked according to their significance and power in the local implementation process.

** The actors' interests in implementing the stated measures are valued as: support (+), neutral (0), resistance (–); the question mark (?) means that the position of the actor is 'not homogeneous'.

Source: Weiss (1998: 123).

According to the principles of their 'forest-technical system' the local TACS officer usually proposes afforestation wherever ecologically feasible but rarely pursues this goal when faced with strong opposition from local actors. However, other interest groups or publics, possibly critical of unsustainable land uses and thus able to strengthen the position of the TACS, are not included in the decision-making process (Weiss, 2001).

With regard to protective measures, the results of the empirical study underlying this chapter are the following: 'technical' projects are usually preferred over afforestation projects; only minor afforestation areas are included in projects; and the success of afforestations is often poor because they are not properly protected against damage by animal browsing etc. The lesson from this analysis is that the distribution of power and interests among actors included in the policy network strongly determines the policy outcome.

Bias in Cost–benefit Analysis

In Austria, over one thousand million ATS are invested in prevention measures against torrents and avalanches every year. On average 80% of the project costs are subsidised. The remainder is borne by the beneficiaries, e.g. municipalities, road and railway agencies, etc. As a prerequisite for granting subsidies for natural hazards prevention projects in Austria, cost–benefit analyses (CBA) have to show a positive result. The CBA as applied within the TACS, however, reflects neither the economic goals of the ministry of finance nor the official principles of the TACS.

Although claiming to be 'objective', CBA falls short in this regard both theoretically and pragmatically. CBA as applied in practice is not scientifically sound with respect to economic theory because different value categories are simply aggregated.[3]

It also does not pragmatically reproduce the formal goals of the TACS. The positions considered within the CBA framework do not adequately reflect the official principles of the TACS. Principles formulated by the agency state that (a) the protection of existing facilities should receive priority over the development of new ones, (b) the protection of lives is more important than the protection of material values, and (c) developments should not be supported that result in an increase in future needs for prevention measures.[4] There are positions included in the CBA that support projects that are not favourable according to these principles. Land that might be developed due to the project in question is evaluated based on its expected future value (estimated rise of the property price). This value is usually several times higher than the estimated expected damage of existing houses (the value of the building is depreciated and the damage is reduced according to its probability). Furthermore, the CBA itself does not consider human lives. The hazard zone plans that indicate the number of houses possibly affected by natural disasters serve as a database for the consideration of possible deaths. However, the consideration of potential fatalities does not take place formally. As a result, there is a wide spectrum of possible decisions. The agency does not want its discretion to be restricted by a standardised procedure.

The existing political system undermines the CBA in another way: by using regional economic arguments, even projects that do not produce a positive cost–benefit relation might be realised. The project is then called a contribution to the economic development of disadvantaged regions. Most projects, however, are realised in prospering regions, e.g. regions with tourism. As a result the natural hazards protection policy does not restrict economic development in mountain regions, it supports it. Growing economic activities in these regions means growing threats and an increasing need for protective measures.

The policy evaluation shows that in practice the TACS does not accept CBA as a decision tool at all. Decisions on what projects will be realised are not based on a cost–benefit analysis; the CBA is elaborated when the decision has already been taken and the measures are already planned in detail. Factors affecting the practical decision of the local officer besides the estimated hazard are (a) the interest and the financial capacity of the applicant community, (b) public attention, (c) informal interests of the agency, and (d) the regional economic development or political priorities on a provincial level.

In the case of CBA, we see clearly that central control instruments are easily corrupted by local authorities if these instruments do not support their informal goals.

Operative Deficits of Hazard Zone Planning

Formally (according to the Forest Act), there are two purposes of the hazard zone plans: (a) to prevent the development of hazard-prone areas for housing etc., and (b) to provide comprehensive data on the hazardous processes as a basis for systematic planning of prevention. Informally, the TACS uses the plans to argue for the elaboration of prevention projects in villages/towns if buildings are situated in 'red zones' (high hazard zones). In a second step, the plans serve in evaluating the achievements of the service in terms of their ability to provide for the safety of the community. The municipalities, who have to cover a share of the project costs, want to 'see results'. A higher standard of safety is visible in technical constructions and furthermore in changed hazard zones: after the completion of a project, 'red zones' are reduced in the maps. What counts politically is not the real reduction of the possibility of a hazard, but the reduction of hazard zones on the map. With the elimination of 'red zones' these areas are available for economic development. The value of these properties may increase dramatically, especially if situated in tourist regions. As shown above, the expected rise of property values is a prominent element within the CBA prepared for torrent and avalanche projects. The CBA is often only positive when these economic effects are included (Weiss, 1999).

The problem with this procedure is that risks are partly ignored. The hazard zone maps provide only restricted information; the hazard assessment refers to a 'standard or reference event' (*Bemessungsereignis*). This reference event is a disaster of a magnitude that is statistically expected every 100 or 150 years. The remaining risk of events with higher impact, the uncertainties of the assessment, as well as the uncertainties of the construction measures are not considered in the assessment and are not shown in the map. Also, the construction layout is designed to deal with the 'standard event' only. If new houses are erected in former 'red zones', now protected by technical constructions, they are still threatened by remaining risks and uncertainties. There are examples in Austria where newly built houses were damaged by events that were more dramatic than 'expected'. In other cases the prevention measures failed (Aulitzky, 1968, 1970, 1996).

One hundred years ago the reference event for prevention constructions was an event with a 30 – year probability of recurrence; 50 years ago this standard was 50 years. Todays measures probably do not meet future expectations. The development of areas protected by technical constructions inevitably results in future damage. The protection gained by the constructions is given away if new houses are erected. Such a policy enlarges the development area but does not reduce overall hazard. In his detailed analysis of the economics of avalanche protection, Wilhelm (1997, 1999: 54) concludes that from an economic viewpoint hazard zones should only be revised and land developed if no alternative areas are available and if the permanent maintenance of the constructions is guaranteed. At least for Austria, that even the most Alpine villages still have safe land without the need for protective measures.

In terms similar to the 'risk circle', described by Sieferle and Mueller-Herold (1996), one could also speak of a 'construction circle", where more and more prevention structures are built, but total safety does not increase (Weiss, 1999: 265). The situation, of course, is worsened in the case of climate change.

4 CLIMATE CHANGE: POSSIBLE EFFECTS AND ADAPTATION

Dramatic Effects on the Safety from Natural Hazards

The effects of changing climate conditions (increased precipitation, higher frequency and higher intensity of heavy rains) on the safety of Alpine areas would be dramatic. These effects are even higher in our cultural landscapes and with protective structures. A considerable portion of infrastructure and settlements already relies on technical constructions for hazard control. As these constructions are usually set up to cope with certain magnitudes, all events with higher impact cause damage – and very often especially high damage. Aulitzky (1968, 1970, 1996), Jaeggi (1996), Kienholz (1996) and Weinmeister (1995, 1996) report on bad experience with certain technical structures and warn that planning for 100% control of natural processes can only be fallacious. Remaining risks and uncertainties should be included in the prevention concept as such.

Their reports include examples of failing structures, causing higher damage than would have occurred if no prevention measures had existed at all. Houses are built in areas that would have never been developed in the absence of artificial structures, and the destruction of such structures may result in debris loads accumulated over decades. The amount of transported debris in a mud-flow is smaller if the event occurs more frequently, according to the natural process. Different technologies for the prevention of mud-flows behave differently in the case of an extreme event: (a) stable check-dams (pools) that hold back a certain amount of water and debris (*Auffangbecken*) at least diminish mud-streams, even in a big event; (b) sequences of check-dams that are expected to stabilise the torrent beds against erosion in the case of high water or mud-flow (*Staffelungen*) are susceptible to damage in the case of a severe event. In such a case the whole sequence of dams may break down (domino-effect) and the debris material that has been accumulated over time rushes down in one event. This means that prevention measures in the past can produce increased damage in the future. This threat is worsened if climatic conditions favour more extreme weather situations.

Dynamic Hazards Zone Planning?

If climate conditions change, obviously the informational basis of the hazards prevention system has to be revised. This problem has already been discussed in the past with regard to the phenomenon of forest decline (*Waldsterben*) and similarly concerning the condition of protective forests and the functions of avalanche protection measures. In terms of forest decline, it was said that forests would lose their capability of buffering precipitation, the consequence being higher floods in the case of heavy rainfall. The logical consequences would have been to enlarge the hazard zones (*Gefahrenzonen*) as well as the declaration of protective belts of forest (*Bannwaelder*). In practice, however, no acceptance was gained for these attempts because they restrict use options.

Similar suggestions were brought up locally in cases of damaged protection forests. There are many cases of poorly managed forests that grow old but lack regeneration due to grazing or overpopulation of game. As they were expected to gradually lose their protective function, protection forest restoration measures (*Schutzwaldsanierung*) were planned but not always implemented. In order to reach an effective resolution of the conflicting interests, the authorities used the argument that hazard zones would have to be enlarged should no action be taken to manage forests properly. So far, no hazard zones have been revised but forests are being restored by means of public funds.

Enlarging red zones (no development allowed) was discussed repeatedly after severe avalanche catastrophes. Ordinances of the ministry, striving for the extension of red zones, met with resistance from local authorities and were withdrawn after communal protests. After the catastrophic avalanche in Galtuer, in 1999, the ministry is again trying to change guidelines in order to enlarge red zones. For the time being it is not possible to say whether this latest attempt will be successful.

The functionality of preventive measures against avalanches is by no means certain. Avalanches may even start from the area provided with preventive structures. Therefore, local authorities of the TACS are often rather cautious with respect to the revision of hazard zones. As stated, after completion of protective measures, hazard zones are revised by the

TACS and the protected areas are available for development. In the case of avalanche protection, however, the TACS argues that full functionality of the measures will only be reached after new forests have been established in these areas (if the areas are suitable for afforestation). However, it shows that this is hard to argue in front of a mayor eager for local tourist or housing development.

The three examples of unsuccessful attempts to enlarge the red zones show that chances for a stricter policy in preparation for climate change are very limited. Furthermore, even the problem of climate change itself has not yet been discussed thoroughly within the TACS.

A Comprehensive System for Natural Hazards Prevention

To date various authors have presented proposals for advanced systems of natural hazards prevention ('sustainable', 'integrated', 'proactive', 'ecosystem-based' prevention). Weinmeister (1988, 1995, 1996) formulates goals for sustainable prevention of natural hazards taking technical, economic, geomorphologic and landscape ecology aspects in consideration. In particular he draws attention to the future cost of maintenance work in technical construction (and to the high energy consumption here necessary). Also, referring to the limited effects of such structures, he emphasises the importance of prevention-oriented land use, zoning, early warning and evacuation plans.

Jaeggi (1988, 1996) also suggests a combination of different tools for coping with the risks resulting from the limits of prevention: protective structures and secondary structural measures, hazard zoning, warning systems and disaster management. He in particular points out that the conventional system tends to hide remaining risks and calls for their deliberate acceptance rather than suppression.

Kienholz (1993, 1996) presents active and passive measures for hazards prevention. He puts emphasis on the remark that prevention systems should not falsely be oriented towards 100% security because this is not realistic. He reports that advanced prevention systems can be developed if the planner is aware of remaining uncertainties. If they are considered they can be controlled to a certain extent through proper structure design and by informational measures as stated above.

The main conclusion from the analysis presented here is that institutional and procedural reforms are necessary (Weiss, 1999: 262ff.). Kienholz (2000), for instance, shows that in recent catastrophic floods in Switzerland severe failures occurred in the communication system, resulting in delayed warning and evacuation. The same conclusions are drawn by Mueller and Zimmermann (1997) who have studied different technical and natural catastrophes in Switzerland. For successful disaster management they point out three important features: communication, co-operation and a crisis-handling committee. Crises are managed 'politically', as they put it.

My own conclusions are that institutional innovations are much more promising than increased investments in technical solutions. The following important proposals can be drawn from this analysis (Weiss, 2000b: 36):

- Avoid the 'risk circle' or 'construction circle': New buildings should not be built in areas depending on the protection of technical structures. Hazard zones should not be reduced after building protective structures. So far only the Tyrolean Land Use Planning Act (*Tiroler Raumordnungsgesetz*) provides that land is not developed closer to hazard areas.
- Reassess levels of acceptable risk in hazard zone plans and develop disaster management plans: Risks beyond the boundaries of hazard zones and in areas below protective structures should be made transparent and precautions should be planned to cater for risky situations (e.g. periods of high risk of avalanches) or catastrophes (evacuation plan, etc.).
- Reform avalanche committees: Committees should comprise of members from outside the community concerned. At present they are biased in their assessment of risk situations because of their members' involvement in local economic interests. Detailed guidelines are needed for the work of the committees. Their decisions, in case of catastrophes, must be subject to follow-up examinations.
- Refine subsidy policies, and raise financial contributions from beneficiaries: Subsidies for protective measures that primarily benefit economic activities should be significantly lower than measures protecting existing settlements. Land-owners that benefit from preventive measures should contribute to the cost. This leads to a more efficient distribution of financial aid, helps to prevent speculation with hazard-prone land and raises awareness among the persons affected.

These measures seem necessary anyway, but even more so when the increased risk of hazard as a result of climate change is considered. In such a case, additional measures would be needed: All hazard zone plans would have to be revised, and existing protective measures would have to be re-examined with regard to their impact behaviour in the case of extreme events. In addition, further research would be needed to estimate the possible effects and consequences for prevention systems. Additional training programmes and extension services would also be important.

5 SUMMARY: NEED FOR INSTITUTIONAL CHANGE

The evaluation of the existing natural hazards prevention policy shows that although the official goals aim at sustainable, integrated and pro-active hazards control, implementation is highly inadequate. The principles of sustainable hazards control are undermined by short-term economic interests. At present it appears that natural hazards control serves not so much the reduction of risk as the demand for economic development: Typically, risk reduction is coupled with developing land for economic or settlement purposes. Subsidies for risk reduction can be seen as subsidies for land development in hazard-prone areas. If touristic development increases in Alpine regions, any risk reduction resulting from hazard control may in fact be outweighted by the increase in activity in mountain areas.

In this situation it is questionable whether the present system can cope, particularly when the growing threats associated with climate change are taken into consideration. Raising awareness within the public administration seems to be possible. However, there still remains the problem of how to put these goals into action. Government institutions are not likely to be strong enough to implement them against powerful local interests. A policy change is only to be expected if an institutional change takes place. Such a change would depend on high political pressure, resulting for instance from greater public awareness in the wake of a catastrophe. Further strong support could be expected from the economy if important infrastructure facilities are threatened.

A policy change is not totally unrealistic – the Hazard Zones Plan and the Tyrolean Land Use Planning Act are positive examples. Hazard zone planning, in principle, is an effective instrument to direct development activities. It is also to be expected that once the regulations in the Tyrolean Land Use Planning Act are suitably adapted and adopted, the 'risk circle' will become avoidable. However, these examples also make it clear that policy innovation takes time, and that there is always the threat that regulations are corrupted in implementation.

In summary, it can be said that institutional changes are necessary to redirect natural hazards policy – changes of actors and changes of procedures. Proposals for a reform of the risk prevention system exist, but political pressure is needed if such innovations are to be realised.

NOTES

1. The research study on 'Evaluation of policy instruments for natural hazards prevention in Austria' was financed by the Ministry of Agriculture and Forestry, and was carried out at the Institute of Forest Sector Policy and Economics, Agricultural University Vienna. It was the basis for the author's dissertation (Weiss, 1999).
2. Recently a study about the options of public participation in torrents and avalanche control projects in Austria has been carried out (Rappold, 2001). This topic is considered in ongoing studies in Europe ('Public participation in mountainous ecosystem management', EFI; EU concerted action 'Multifunctional forest management in mountain regions – evaluation of silviculture and policy means', MUFOMA).
3. Guidelines ('Kosten-Nutzen-Untersuchung'), BMLF (1978).
4. For example, guidelines 'Hinderungsgruende', BMLF (1980).

BIBLIOGRAPHY

Aulitzky, H., 1968: Analyse der Schadensursachen von Unwetterkatastrophen zum Zweck der Vorbeugung. *Oesterreichische Wasserwirtschaft* 20, H. 5/6 und 7/8.

Aulitzky, H., 1970: Der Enterbach (Inzing in Tirol) am 26. 7. 1969: Versuch einer Analyse eines Murganges als Grundlage fuer die Neuerstellung einer zerstoerten Wildbachverbauung. *Wildbach- und Lawinenverbau* 34, H. 1, pp. 31–66.

Aulitzky, H., 1996: Beispiele vermeidbarer Hochwasser- und Lawinenschaeden. In: *Naturkatastrophen – hausgemacht?* Conference Proceedings. Wien: Oesterreichische Gesellschaft fuer Oekologie, pp. 25–55.

BMLF, 1978: *Kosten-Nutzen-Untersuchung.* Wien: Bundesministerium fuer Land- und Forstwirtschaft.

BMLF, 1980: *Richtlinien Hinderungsgruende.* Wien: Bundesministerium fuer Land- und Forstwirtschaft.

Compare, Alexander, 1995: *Risikokommunikation als Machtproblem.* Europaeische Hochschulschriften 22. Frankfurt/Main: Lang.

Conrad, J., 1989: Risiko, Ritual und Politik. In Schmidt, M. (ed.), *Leben in der Risikogesellschaft. Der Umgang mit modernen Zivilisationsrisiken.* Karlsruhe: Mueller, pp. 179–204.

Daele, Wolfgang van den, 1994: *Technikfolgenabschaetzung als politisches Experiment: diskursives Verfahren zur Technikfolgenabschaetzung des Anbaus von Kulturpflanzen mit gentechnisch erzeugter Herbizidresistenz.* Berlin: WBZ.

Dienel, Peter C., 1997: *Die Planungszelle: der Buerger plant seine Umwelt; eine Alternative zur Establishment-Demokratie.* Opladen: Westdt. Verl.

Dietz, T.M. and Rycroft, R.W., 1987: *The Risk Professionals.* New York: Sage.

Downs, A., 1967: *Inside Bureaucracy.* Boston: Little and Brown.

Edelmann, M., 1976: *Politik als Ritual.* Deutsche UEbersetzung. Frankfurt/Main: Campus.

Fischhoff, B., S. Lichtenstein, P. Slovic, S.L. Derby and R.L. Keeney, 1981: *Acceptable Risk.* Cambridge/New York: Cambridge University Press.

Glueck, P., 1976: *Die Rolle der Verbaende in der theoretischen Forst- und Holzwirtschaftspolitik.* Wien: Habilitationsschrift an der Universitaet fuer Bodenkultur Wien.

Hollenstein, K., 1995: Analyse und Bewertung von Risiko und Sicherheit bei Naturgefahren. In: *Schweizer Zeitschrift fuer Forstwesen* 146/9, pp. 687–700.

Jaeggi, M., 1988: Sicherheitsueberlegungen im Flussbau. *Wasser Energie Luft* 9, pp. 193–197.

Jaeggi, M., 1996: Gedanken zu den Grenzen der Vorbeugung. *Oesterreichische Wasser- und Abfallwirtschaft* 48, H. 5/6, pp. 179–186.

Jaenicke, M., 1986: *Staatsversagen: die Ohnmacht der Politik in der Industriegesellschaft.* Muenchen, Zuerich: Piper.

Kienholz, H., 1993: Naturgefahren – Naturrisiken im Gebirge. In: *Forum fuer Wissen 1993*, pp. 7–22.

Kienholz, H., 1996: Naturprozesse und Siedlungsentwicklung. Paper presented at *FORUM ALPIN 1996*, Chamonix, France.

Kienholz, H., 2000: Hazard assessment, communication and implementation. Paper presented at CALAR Conference *'Living with natural hazards'*, Vienna, Austria.

Krott, M., 1990: *Oeffentliche Verwaltung im Umweltschutz. Ergebnisse einer behoerdenorientierten Policy-Analyse am Beispiel Waldschutz.* Wien: Braumueller.

Luhmann, N., 1966: Politische Planung. In: *Jahrbuch fuer Sozialwissenschaft.* Goettingen.

Luhmann, N., 1993: Risiko und Gefahr. In: Krohn, G. und Kruecken, W. (eds.): *Riskante Technologien: Reflexion und Regulation. Einfuehrung in die sozialwissenschaftliche Risikoforschung.* Frankfurt/Main: Suhrkamp, pp. 138–185.

Mayntz, Renate, 1985: *Soziologie der oeffentlichen Verwaltung.* 3. ueberarb. Auflage. Heidelberg: Mueller.

Mazur, A., 1985: Bias in Risk-benefit Analysis. *Technology in Society* 7, pp. 25-30.

Mueller, U. and W. Zimmermann, 1997: *Katastrophen als Herausforderung fuer Verwaltung und Politik.* Schlußbericht des NFP 31. Zuerich: Vdf.

Nelkin, D. and Pollack M., 1980: Problems and Procedures in the Regulation of Technological Risk. In: Schwing/Albers (eds.), pp. 233–248.

Perrow, C., 1984: *Normal Accidents. Living with High-risk Technologies.* New York: Basic Books. Dt.: *Normale Katastrophen. Die unvermeidbaren Risiken der Großtechnik.* Frankfurt/Main and New York: Campus.

Peters, B.G., 1984: *The Politics of Bureaucracy.* New York and London.

Pregernig, M. and Weiss G., 1998: Forest Policy in Austria: Policy Making by the Sector for the Sector. Paper presented at the IFF (Intergovernmental Forum on Forests) European Regional Meeting on the Underlying Causes of Deforestation and Forest Degradation in Europe, 28–29 October, Bonn.

Prittwitz, V., 1990: *Das Katastrophenparadox. Elemente einer Theorie der Umweltpolitik.* Opladen: Leske+Budrich.

Rappold, G., 2001: *Buergerbeteiligung im Naturgefahrenmanagement.* Wien: Dissertation an der Universitaet fuer Bodenkultur.

Sabatier, P.A., 1993: Advocacy-Koalitionen, Policy-Wandel und Policy-Lernen: Eine Alternative zur Phasenheuristik. In: Héritier, A. (ed.), 1993: *Policy-Analyse. Kritik und Neuorientierung.* Politische Vierteljahresschrift (PVS), Sonderband 24/1993. Opladen: Westdeutscher Verlag. pp. 116–148.

Scharpf, Fritz W., 1997: *Games Real Actors Play. Actor-centered Institutionalism in Policy Research.* Boulder, Oxford: Westview Press.

Schwing, R.C. and Albers, W.A. (eds.), 1980: *Societal Risk Assessment. How Safe is Safe Enough?* New York: Plenum Press.

Shrader-Frechette, K.S., 1995: Evaluating the expertise of experts. *Risk* 6(1), 99.1–9.

Sieferle, Rolf Peter und Ulrich Mueller-Herold (1996): UEberfluß und UEberleben – Risiko, Ruin und Luxus in primitiven Gesellschaften. In: *GAIA* 5, no. 3–4, pp. 135–143.

Starr, C., 1969: Societal Benefit versus Technological Risk. What is Our Society Willing to Pay for Safety? In: *Science* 165, pp. 1232–1238.

Starr, C., 1980: Introduction to the session. In Schwing/Albers (eds), pp. 1–4.

Weinmeister, H.W., 1988: Oekonomie und Oekologie am Beispiel des Schutzwasserbaues (landschaftsoekologische und volkswirtschaftliche Grenzen des technischen Schutzwasserbaus). Proceedings of the conference *Interpraevent* 1988, vol. 4, pp. 339–376.

Weinmeister, H.W., 1995: Ideas and Proposals for a Sustainable Protection from Torrents and Avalanches. Proceedings of the International Sabo Symposium, Tokyo, Japan, pp. 329–333.

Weinmeister, H.W., 1996: Nachhaltige Entwicklung der Bergregionen. Grundgedanken fuer einen vorbeugenden Schutz vor Katastrophen. Paper presented at Interpreavent 1996, Garmisch-Partenkirchen, Germany. Tagungspublikation, vol. 2, pp. 181–190.

Weiss, G., 1998: Evaluation of Mountain Forest Policy in Austria. In: Glueck, P. and Weber, M. (eds., 1998), *Mountain Forestry in Europe – Evaluation of Silvicultural and Policy Means.* Publication Series of the Institute for Forest Sector Policy and Economics 35. Universitaet fuer Bodenkultur Wien, pp. 107–136.

Weiss, G., 1999: *Die Schutzwaldpolitik in Oesterreich. Einsatz forstpolitischer Instrumente zum Schutz vor Naturgefahren.* Wien: Proceedings of the Institute of Forest Sector Policy and Economics 39. Also: dissertation at the Universitaet fuer Bodenkultur Wien.

Weiss, G., 2000a: Perceptions of risks in mountain forests. In: Price, M. and N. Butt (eds.), *Forests in Sustainable Mountain Development: A State-of-knowledge Report for 2000.* Oxon: CABI, pp. 359–366.

Weiss, G., 2000b: Aus 'Galtuer' gelernt, oder? In: *RAUM* 39/00, pp. 34–36, Wien.

Weiss, G., 2001: Administrating Risk – A Social Science Perspective on Natural Hazards Prevention Based on an Austrian Case Study. In: Gadow, K. v. (ed.), *Risk Analysis in Forest Management.* Dordrecht, Boston, London: Kluwer, pp. 171–200.

Wilhelm, C., 1997: *Wirtschaftlichkeit im Lawinenschutz: Methodik und Erhebungen zur Beurteilung von Schutzmaßnahmen mittels quantitativer Risikoanalyse und oekonomischer Bewertung.* Davos: Mitteilungen des Eidgenoessischen Instituts fuer Schnee- und Lawinenforschung; 54. Zugleich Dissertation ETH Zuerich.

Wilhelm, C., 1999: Naturgefahren und Sicherheit der Bevoelkerung im Gebirge – oder: von der Schicksalsgemeinschaft zur Risikogesellschaft. In: *Forum fuer Wissen* 1999, 2. Birmensdorf: Eidgenoessische Forschungsanstalt fuer Wald, Schnee und Landschaft, pp. 47–56.

Wolf, R., 1986: *Der Stand der Technik. Geschichte, Strukturelemente und Funktion der Verrechtlichung technischer Risiken am Beispiel des Immissionsschutzes.* Opladen: Westdeutscher Verlag.

Wynne, B., 1992: Risk and Social Learning: Reification to Engagement. In: S. Krimsky and D. Golding (eds.), *Social Theories of Risk.* Westport, CT: Praeger, pp. 275–297.

9. Climatological Research and its Possible Contribution to Regional Planning in an Alpine Environment

Wolfgang Sulzer

1 INTRODUCTION

The relationship between climatic basics and human activities in an Alpine environment is crucial. Traditional land use has been influenced by climate (local and global) for many centuries. By exploiting more and more natural resources (e.g. natural or forested areas, etc.) human beings have ventured into very sensitive climatic regions. One example is the intensive use of valley bottoms. Residential and industrial areas have been built up particularly in the last four decades and thus affect the local climate, mainly through the increase of pollutants.

The governmental planning institutions of the Austrian Province Styria use development plans and maps for regional and local planning. The interaction of human and natural activity and climate, especially in residential and urban areas, is integrated into these maps. Such suitable climatological maps are constructed for the whole province (scale 1:50 000) to support governmental institutions in their task of siting industrial and residential areas. After surveying the natural and anthropogenic climate factors for alpine residential areas in section 2, the subsequent section 3 focuses on the methods used to construct climate suitability maps.

2 ALPINE RESIDENTIAL AREAS AND THEIR CLIMATE

Recent urban climatology research has often dealt with the phenomenon of the urban heat island and its meso-circulation of local wind systems. This phenomenon significantly influences the spread of pollutants. A further topic of climate research is the prediction of the local wind system by computer models and the coupling of the results with local emission maps.

This research has fed back on regional development plans and land use regulation. For example, the problems encountered with smog in the winter of 1988/89 in the Styrian capital city of Graz led to the introduction of a regional clean-air plan involving the implementation of additional precautionary measures, e.g. changes in domestic heating systems, and obligatory use of district heating. An extensive analysis of urban climate was carried out before. Most recently, a model for the simulation of pollutant transport in Graz was developed by Sturm et al. (1999) and Almbauer et al. (2000). With this simulation method it is possible to predict the dispersion of different pollutants within an urban area.

The specific demands of regional planners and their interaction with scientists then led to the creation of climate suitability maps. Before their use and development is discussed in detail in section 3, we shall survey the general factors of the alteration of climate in mountainous regions.

2.1 The Alteration of Climate in an Alpine Urban Environment

The diversity of the outward form of the urban climate, which gives each Alpine town its individual climatic variation, derives from different natural and anthropogenic factors. These specific factors influence regional climates. They include:

Natural factors

- *Geographical latitude* The thermal balance is mainly influenced by the angle of the sun and intensity of radiation.
- *Type of land cover* The thermal balance is particularly influenced by the soil type and land cover (albedo).
- *Altitude* Climatic elements vary according to height above sea level. Generally speaking, the temperature will decrease and the precipitation will increase with altitude, but there are exceptions. Within inversions (especially in Alpine valleys) the temperature tends to increase with height above ground.
- *Effects caused by the shape of the landscape* The topographic situation is of prime importance for specific climate characteristics in Alpine regions. Thus, the climatic variations of a flat area near the sea are different to those of an Alpine basin. Flat country is excellently aerated because of higher wind velocity, extenuation of daily and yearly temperature variation, more periods of sunshine. For areas in coastal regions, the presence of specific land and sea wind

systems also affects climate conditions. Mountainous areas and Alpine forelands with valleys and basins are characterised by lower wind velocity, local wind systems, the effects of foehn, an increase of fog, accumulation of cold air, higher frost rate, higher variation of daily and yearly temperature and often a higher load of pollutants and specific precipitation variations caused by the local relief. On slope areas, however, less fog develops and there is a balanced daily temperature and a smaller pollutant burden.

- While the above effects are a reflection of the natural climate, humanmade alterations in land-cover systems lead to specific variations in the existing local climate:

Anthropogenic factors

- *Sealed areas* The density and spatial expansion of settlements influence climatic conditions e.g. through their impact on pollutants from private houses, industrial factories, traffic and power stations.
- *Other land-cover changes* such as clearing activity, afforestation, irrigation, drainage and cultivation alter the local climate too.

The elements and processes of urban climate produce specific alterations in the local climate including:

- *Formation of urban heat islands* through the occurrence of different thermal balances and the production of thermal radiation (especially during the night).
- *Increased thermal radiation* during the summer, especially as a result of reduced night cooling.
- *Local wind systems* will be influenced by sealed areas, depending on the height and alignment of the houses.
- *Inversion* with variable thickness of the potential mixture sphere.
- *Less exchange of fresh air* because of the blocking of local wind systems.
- *Haze build-up* above settlements.
- *More foggy days* due to an increased quantity of condensation kernels.
- *Uplifted fog* during high pressure periods.
- *Industrial snowfall.*
- *Limited natural ventilation.*
- *Fewer days with snow cover.*
- *Modified effects of precipitation.*
- *Reduced air quality.*

Alpine towns, which are situated within sheltered valley bottoms and basins, are less exposed to the influence of weather conditions coming from beyond the Alps. This situation leads to a more continental climate with increased seasonal diversity, especially in terms of temperature and precipitation. There is a marked lack of strong winds in winter, which favours the development of inversions and valley fog. Due to the topographical situation of Alpine valleys and basins, local wind systems will be well developed. The percentage of periods without any clouds and strong winds is higher in the central part of the eastern Alps than in the northern part and forelands. They reach their maximum in autumn with 60–70% of all days (minimum in spring with 40–50%; these periods are interrupted by many local thunderstorms during the summer and by long periods of uplifted fog during the winter).

The diversified structure of housing (density, sealing rate, structure and alignment of roofs) leads to differences in temperature, especially between daytime and night time readings (see Table 9.1).

Table 9.1: Temperature within different housing structures in Graz

Type of structure	Density	Temperature during day (°K)	Temperature during night (°K heating)
Old historic town Compact structure, high sealing rate	2.0	Differences caused by local climate conditions (e.g. shadow, heating)	3–5
End of 19th century Uniform floors, partially open patios	1.5–2.0		2.5–4
Block belt Loose structure, single tower blocks, community building	2.0–2.5		2.5–4
Garden city Mixture of line structure and high, dense, small buildings	0.8–1.5		1.5–3
Villa belt Low density and much greenery	< 0.5		0.5
Outskirts Low density and mixture with agricultural areas	0.6–1.5		0.5–2
Industrial areas High sealing rate	2.0–3.0		1–3

Figure 9.1 shows an aerial photograph of some parts of an adjacent small valley in the eastern part of Graz and its typical housing, which is transversely oriented to the direction of the valley, and to the slope. This

arrangement results in limited wind circulation. The local wind system is blocked by the housing (2), so that no or less cold fresh air can flow into the valley bottom. (1) represents the valley bottom, (3) housing arranged transversely to the valley direction and (4) housing transverse to the slopes. The arrows show the local wind system during the night and at early dawn. This situation is typical for naturally undulating areas normally found within an Alpine environment.

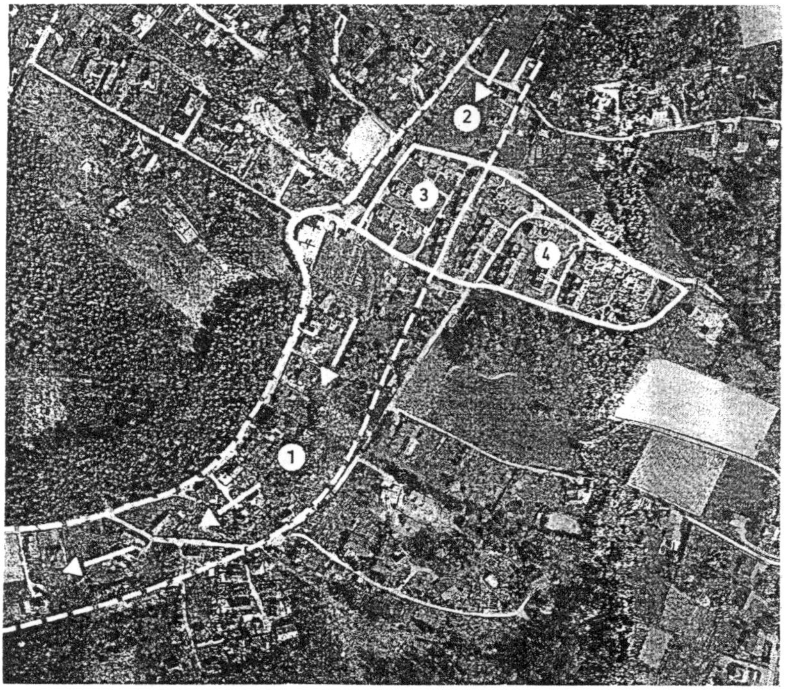

Source: Lazar et al. (1994), explanation in text.

Figure 9.1: Typical climatic situation in the eastern part of Graz

Concerning the modifications of the climate caused by human activities on the one hand and by the influence of the topographical situation on the other, it is possible to generate climatope maps and, as a further step, climate suitability maps.

3 CLIMATE SUITABILITY MAPS OF STYRIA – SOME EXPERIENCES

The local and regional governments use climate suitability maps for planning purposes. The Styrian government produces such maps for the whole province. These 1:50 000-scale user-friendly maps relate planning-based information to climatic suitability. Each map contains a base map and two cover maps (transparent layer maps for wind-movement system and climate suitability). For regional planning purposes and for questions relating to air pollution, the maps have already proved themselves to be an important decision aid. The maps are prepared in co-operation with ARGE LOESS and the Institute for Geography and Regional Science (University of Graz) for the Styrian Government (Bader et al., 1993).

The eastern tertiary area in the foreland of the Alps is very well documented. The recently available maps of the Alpine regions focus on widespread, populous valleys (Mur – Muerz Valley and Enns Valley in the vicinity of Liezen).

The first edition of the maps was made available in analogue form. The latest editions are digitised and have been integrated into a government Geographic Information System (GIS) to advance their application in planning purposes (the latest information about available maps can be found at: http://www.stmk.gv.at/umwelt/luis/Naturraum/STARTSEITE/index.htm.

The output map is also to be integrated into the official regional and local plan (governmental law). This will influence village and regional development in that certain constraints will be placed on housing schemes and on the establishment of new industrial zones. Even the type of heating system can be influenced by the plans. For example, residential houses in small valleys with a typical local wind system, influencing the transmission of pollution to other settlement areas, must use non-conventional heating systems, not single fossil fuel heating units.

3.1 Database and Recording Methods

The basic requirement for generating these maps is detailed knowledge of climate indicators, especially of those found in an Alpine environment. The natural climatic input (temperature, local wind system, fog, etc.) must therefore be investigated first. As a next step the influence of human activities on the variation of these climatic elements (variation of temperature in residential areas, influence of infrastructure on wind system, etc.) must be detected and mapped by fieldwork. It is very important during this phase to clarify and make explicit how the requirements of regional planning can best be matched to the climatic investigation. Here, intensive collaboration and interaction between climate researchers, planners and government greatly improves the value of the studies.

The recording methods are divided into three steps: (1) analysis of available databases recorded by the Central Institute for Meteorology and Geodynamics in Vienna and the Hydrological Section of the Styrian Government; (2) installing a special investigation network (including climate recording stations) in the investigation area within one or two years and; (3) survey of local climate situations, e.g. measuring local differences of temperature and the influence of the local wind system using fieldwork; measuring the vertical temperature and wind system by launching balloons and analysing thermal, infrared, remote sensing images.

3.2 Structure of the Suitability Maps

The climate suitability maps consist of a basic topographical map, in which the different climatopes of the areas under investigation are visible. The near-ground fog situation and wind system are included as an additional information layer. The 'Cover Map I' includes information about the wind system between ground level and 50–100 m and the 'Cover Map II' sums up details about building suitability, particularly with respect to land utilisation (industrial areas, residential areas, etc.).

Figures 9.2 to 9.4 show parts of the suitability map for the Alpine Mur Valley in the vicinity of Bruck with Basic Map, and Cover Maps I and II. The different climatopes are described by the alteration of the following factors:

- temperature structure;
- amount of inversion days;
- fog structure;
- velocity and direction of wind.

Cover Map I shows the direction (lines) and velocity of local wind systems with levels of shading (from 0.5 m/sec. – white, to larger than 5 m/sec. – black), an overlay of different systems (lines with small pikes).

Cover Map II represents the suitability map with a valuation for natural airing, expressed by a 'D' and numbers (1 = best, 5 = worst; superscripts show the situation 50–100 m above ground, subscripts the situation near the ground). The 'I' represents the amount of the temperature inversion (1 = best, 5 = worst) near the ground, 'N' the foggy situations

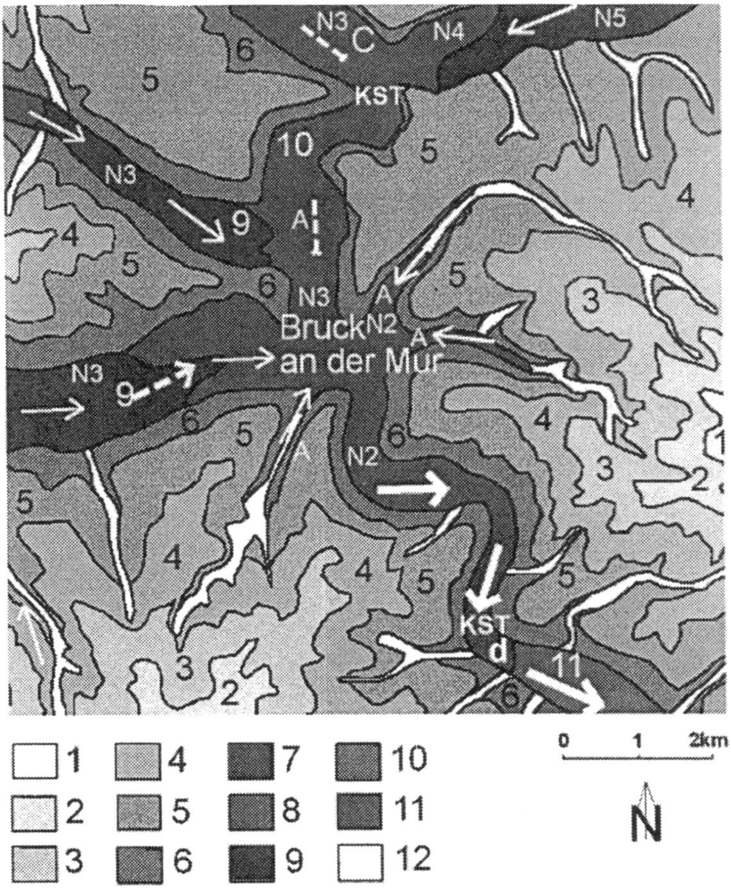

Key to basic climatope map:	C: Valley bottoms
A: Peaks, ridges and slopes	8: Fog abundant areas in main valley
1: High ridges above 1 400 m	9: Main valleys (general)
2. Moderate slopes above 1 200 m	10: Low-favoured valley bottoms
3: Lower ridges between 900 m–1 000 m	11: Favoured valley bottoms
4: Higher uplifted fog zone (800–1 000 m)	12: Steep valleys
5: Lower uplifted fog zone (600–800 m)	(C = calm areas, arrows = different
B: Terraces	velocity, d = pipe wind effect,
6: Thermally less-favoured terraces	N = different amount of foggy days)
7: Lower terraces	

Figure 9.2: Basic map from the climate suitability map for the vicinity of Bruck

with the same 1–5 weighting. ‘G’ (industrial areas) and ‘W’ (residential areas) represent the suitability for these land-use categories (‘+’ means no special limitation from a climatic point of view, no additional sign means no requirements and ‘–’ means not fit for this land-use category from a climatic point of view). ‘FLZB’ (small horizontal lines) signify special areas, where housing should be forbidden, due to their function as an airing area for existing residential regions.

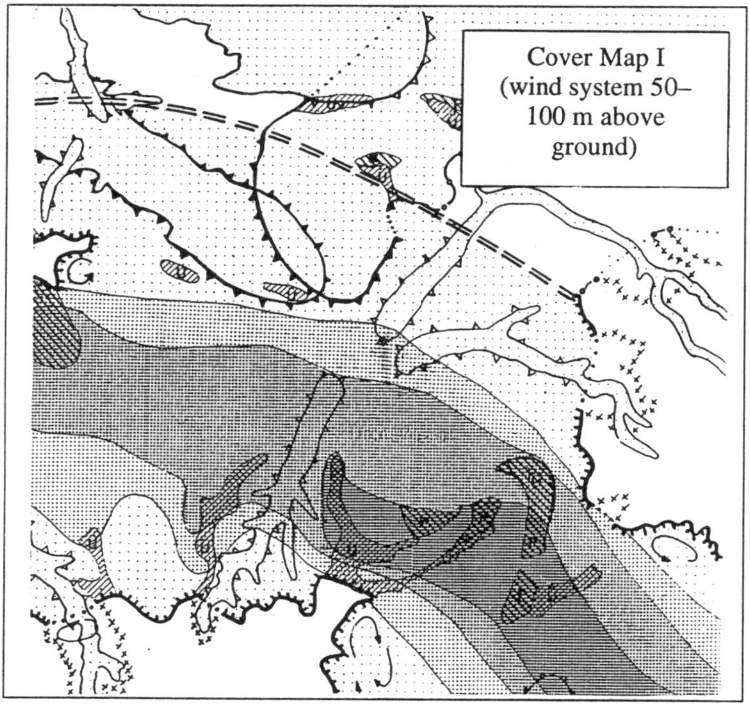

Figure 9.3: Cover Map I from the climate suitability map for the vicinity of Bruck (B/W version, explanation in text)

Detailed information about the structure of the Climate Suitability Maps and their legend symbols can be found in Kuncic and Lazar (1997) or on the homepage of the Government of Styria (http://www.stmk.gv.at/umwelt/luis/Naturraum/klima/KLIMAEIGNUNGSKARTEN/uebersicht.htm and http://www.stmk.gv.at/umwelt/luis/Naturraum/STARTSEITE/index.htm).

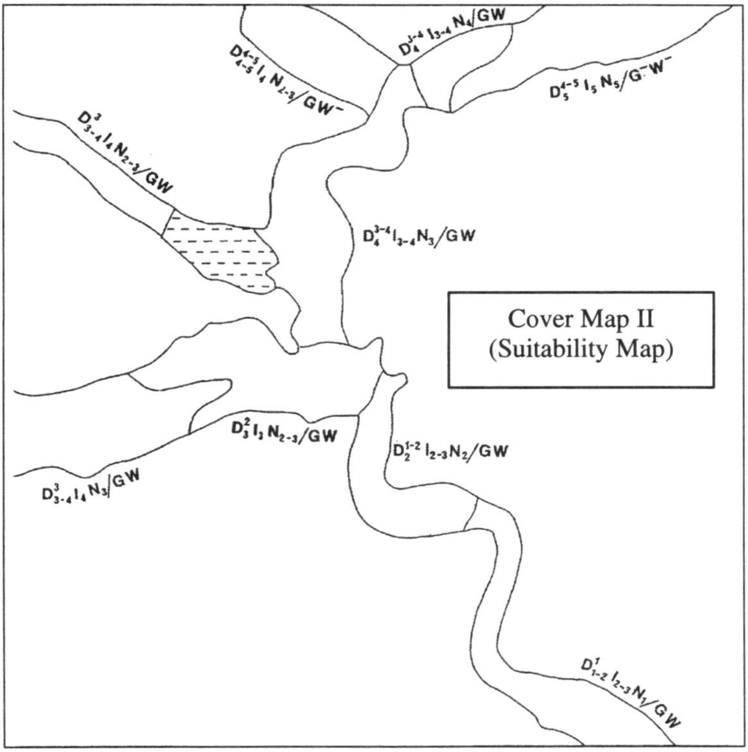

Source: Kuncic and Lazar (1997).

Figure 9.4: Cover Map II from the climate suitability map for the vicinity of Bruck (B/W version, explanation in text)

4 SUMMARY

Many climate suitability maps for the province of Styria have been drawn over the past decade. Most of the valley bottoms and adjacent areas are mapped and the results have been integrated into the government geographical information system.

Recent experience in the application of suitability maps shows that the information from the maps and from the supplementary texts is very important for decision making in the extensive areas of regional and local planning. Such maps are used principally for preparing land-use maps. Many of the building restrictions included in the latter are in fact derived from the respective suitability maps. The province of Styria is exceptional within Austria in taking climatic aspects into account in planning processes. This applies in particular to Alpine environments where the topographical situation leads to specific changes in climate and where, therefore, a very sensitive approach to the influence of human activities is required.

BIBLIOGRAPHY

Almbauer, R.A., D. Oettl, M. Bacher and P.J. Sturm (2000), '*DATE Graz- A Field Study and Modelling Exercise for Urban Air Problems*', Report No. 76, Institute for Internal Combustion Engines and Thermodynamics, TU-Graz.

Bader, Th. A., N. Braun, E. Kuncic, R. Lazar and W. Sulzer (1993), 'Gelaendeklimatologische Untersuchungen in der Steiermark', in *Wetter und Leben*, Heft 3/93, pp. 45–51.

Kuncic, E. and R. Lazar (1997), 'Bisherige Erfahrungen mit Klimaeignungskarten', *Arbeiten aus dem Institut fuer Geographie der Karl-Franzens Universitaet Graz,* Bd. 35, pp. 139–152.

Lazar, R., M.F. Buchroithner and V. Kaufmann (1994), 'Stadtklimaanalyse Graz', *StekGraz des Magistrates Graz – Stadtplanungsamt*, Graz, p. 163.

Lazar, R. and ARGE LOESS (Bader, Braun, Kuncic, Sulzer) (1994), 'Klimaeignungskarte Leoben – Bruck/Mur', im Auftrag der Steiermaerkischen Landesregierung.

Sturm P.J., Sudy Ch., Almbauer R.A., Meinhart J., (1999),'Updated Urban Emission Inventory with a high resolution in Time and Space for the City of Graz', *The Science of the Total Environment*, Vol. 235, pp.111–118.

Sulzer, W. (1997), 'Die klimatischen Verhaeltnisse in Andritz', in: *Bericht zur 2. Naturgeschichtswerkstatt in Andritz*, Graz, pp.12–22.

PART IV

Mitigation

10. The Role of Alpine Agriculture and Forestry in Climate Change Mitigation – a Scenario Analysis

Klaus C. Steininger and

Karl W. Steininger

1 INTRODUCTION

Agriculture and forestry are of great importance in the Alpine region. For example, agriculture and forestry dominate land use in the Alpine region.

It is well known that there is an impact of climate change on agriculture and forestry, i.e. small changes in climate can have significant impact on land-use management. In this chapter, however, we focus on the reverse linkage, namely the relevance of agriculture and forestry as sectors contributing to climate change. The quantitative and qualitative descriptions of this impact are presented and the potential options for climate change mitigation are analysed and quantified.

The chapter is structured as follows. Section 2 provides quantitative indications of the role of agriculture and forestry in the Alpine region. In section 3 the main factors that determine the contribution of agriculture and forestry to climate change are identified as direct and indirect energy consumption, fixation of carbon through soil organic matter (SOM) in different land-use types, and non-energy related production of greenhouse gases. In section 4 an array of indicators for measuring these factors at farm and regional level is presented. On surveying these indicators for 80 alpine farms in five countries the most important factors turned out to be methane emissions and soil carbon fixation. For example, turning arable land into grassland (or forests) can double organic C content and thus fix a considerable amount of atmospheric CO_2.

We then move on to analyse future development opportunities. Section 5 starts with three scenarios from a recent study on Alpine agricultural and forestry development and analyses their implications for the greenhouse gas balance. This section also quantifies the potential greenhouse balance impacts of the most extreme (and unrealistic) scenarios of Alpine agriculture and forestry development in order to derive their maximum potential contribution towards climate change mitigation in the Alpine region. Finally, section 6 summarises the main conclusions.

2 THE ROLE OF AGRICULTURE AND FORESTRY IN THE ALPINE REGION

In terms of their spatial share in land-use agriculture and forestry play a dominant role in the Alpine area. Employing the geographic definition of the Alpine Convention,[1] a quantitative specification of this dominant role is summarised in Table 10.1. Table 10.1 draws on the data of roughly 5 600 communities in seven countries and has been compiled within European Commission R&D-project concerning the evaluation of instruments of the European Union regarding their contribution to sustainable environment and agriculture in the Alps.[2]

Table 10.1: Agriculture and forestry within the total Alpine area

Type of land use	Area (ha)	% of total Alpine area
Total Alpine area (area of communities)	18 097 880	100.00
Total farm-managed area	11 094 130	61.30
Agricultural area in use	2 346 064	12.96
Arable land	574 480	3.17
Permanent grassland	1 589 507	8.78
Permanent cultures (vines, orchards)	45 156	0.80
Agricultural area not in use	505 691	2.79
Forestry	4 952 350	27.36

Source: SUSTALP (2000): Survey of national agrarian statistics.

With the total 'farm-managed' area at 61.3%, almost two-thirds of the Alpine area are influenced by agriculture and forestry. This category includes all areas that are owned or managed by agricultural or forestry farms. Thus, the area owned by these farms but no longer used for agricultural and forestry purposes is included, as are other areas, such as farmyards, water, wasteland and rock. The remainder of the total Alpine area is covered by rock, water, or glaciers, or is used for settlements or transport infrastructure.

Within the agricultural- or forestry-managed area, forest use dominates with a share of almost half. Permanent grassland follows with approximately a seventh. Further main types of land use are arable land, and agricultural areas no longer in use. Permanent cultures, such as vineyards represent less than 1% of the agricultural- and forestry-managed area.

From this data we can conclude that roughly 43% of the Alpine area is directly affected by agriculture and forestry, i.e. land use in this area is a direct result of decisions taken in agriculture and forestry management. When the current or potential contribution of agriculture and forestry to any economic or environmental impact is analysed, only these areas can and will be considered in this chapter.

Table 10.2 provides the same land-use statistics, but with respect to the Alpine area within Austria only. Many of the conclusions in this chapter are restricted to Austrian data. We find the importance of agriculture and forestry to be substantially higher within Austria than in the overall Alpine area. This stems from the stronger importance of forestry and slightly higher importance of permanent grassland.

Table 10.2: Agriculture and forestry within the Austrian Alpine area

Type of land use	Area (ha)	% of total Austrian Alpine area
Austrian Alpine Area	5 437 886	100.00
Total farm-managed area	5 009 309	92.10
Agricultural area in use	780 637	14.30
Arable land	202 300	3.70
Permanent grassland	568 807	10.50
Permanent cultures (vines, orchards)	5 605	0.10
Agricultural area not in use	31 523	0.60
Forestry	2 499 850	45.90

Source: See Table 10.1.

In addition to the land-use importance, further relevant aspects of the existing agricultural structure could also be considered. First, the primary sector plays a significant role in the labour market. Within the total Alpine area the primary sector directly accounts for roughly 5% of employed persons (SUSTALP, 2000; own estimation).[3] Second, there is a large indirect importance as the primary sector produces basic inputs for the processing industry (such as milk for dairy products). Finally, the social importance of the agricultural sector might be considered.

3 IMPORTANCE OF AGRICULTURE AND FORESTRY FOR CLIMATE CHANGE

Analysis has shown that there is a strong link between agriculture and climate. First, small changes in climate can cause important changes in land use. Obviously this relationship is valid for all production processes that depend on abiotic conditions such as temperature or precipitation. For mountainous areas, however, this relationship is more important than for other agricultural areas at lower altitudes. At higher altitudes the vegetation period is shorter, which implies that the variety of agricultural products is comparatively small. Natural conditions in Alpine regions have particular effects on the intensity of land management. For instance, the prevailing lower average temperature and higher precipitation influence the growth rate in grasslands. The number of crops possible in permanent grassland is determined by these natural conditions as well as by altitude, exposition or land inclination. Due to the presence of large topographic variety in the Alpine area these local factors vary strongly, even within one region. Thus, all in all, in Alpine regions land use depends much more directly on prevailing climate and topographical conditions.

In this chapter we focus solely on the second link, the contribution of agriculture and forestry to climate change. The following factors are relevant for the greenhouse effect caused by agriculture and forestry:

- production of greenhouse gases by fossil energy use;
- changes in the carbon stock balance (changes in fixation of carbon in organic biomass);
- production of greenhouse gases from non-energy activities (CH_4, N_2O).

In order to quantify these contributions or important portions thereof we use the following indicators at the farm level:

- energy consumption;
- fixation of carbon as a result of changes in the type of land-use;
- production of greenhouse gases by livestock.

These three points are analysed in more detail below and a specific definition of the respective agricultural and forestry contribution is derived.

3.1 Energy Consumption

For energy consumption three aspects can be distinguished:

1. consumption of renewable energy;
2. direct fossil energy consumption;
3. indirect fossil energy consumption.

Renewable energy
As the share of energy derived from appropriate renewable sources such as hydropower or biomass (given reforestation) increases, the climate change impact of energy use is reduced (see also Chapter 11).

Direct fossil energy consumption
Direct energy consumption covers all energy consumed immediately within the production process for the generation of usable energy (Werschnitzky et al., 1987). The subcategories of direct fossil-based energy consumption are usually fuel (diesel oil, petrol), gas and electricity from thermal production. The climate-relevance is given by the greenhouse gas emissions that arise during the use and/or production of these energy products.

For the quantification of the direct energy consumption two approaches can be distinguished (Diepenbrock et al., 1995): the 'end-energy' approach and the 'primary energy' approach. The first approach does not cover energy losses that occur during the transformation process of primary to secondary energy. It is thus a relatively simple approach for which data is easily available (Werschnitzky et al., 1987; BMELF, 1983). The second approach includes coverage of all energy resources used during all phases of the supply of energy (production and transport included). The data requirement is therefore much higher, as the efficiency of several transformation and transport processes has to be known, as does the energy content in raw materials (Reinhardt, 1993). For purposes of international coverage and uniform data availability we restrict ourselves to the simpler first approach in this chapter.

Indirect energy – consumption
Energy in agriculture is used indirectly by means of industrially produced inputs, such as fertiliser, pesticides, machinery or buildings.

These intermediate inputs to agriculture represent the energy consumption of the industrial processes needed to produce them. The climate-relevance is given by greenhouse gas production during the production of these products.

The quantification of indirect energy consumption is no simple matter since no single method for the energetic description of individual production processes exists. In the literature the following goods are normally listed as relevant: fertiliser, plant protection products, seed, machines, buildings and lubricants (Diepenbrock, 1995). This chapter only considers fertiliser and plant protection products as these two cover the main indirect energy input.

3.2 Changes in the Carbon Household due to Land-use Changes

The basic processes determining the carbon household are the following. Vegetation biomass takes up atmospheric C under release of O_2. In turn organisms use O_2 in respiration and return C to the atmosphere in the form of CO_2. Connecting these two basic processes of global life there are also several sinks and sources where C is bound for a shorter or longer time. One essential sink or source is vegetation biomass; others include 'dead' vegetation (organic components of soil, organic sea sediments, moorlands, fossil fuel reserves), and obviously air, also water, where CO_2 is diluted.

Therefore, in terms of agriculture influencing the carbon household, we can first look at the net primary production of biomass (and thus binding of C) in different vegetation types. Table 10.3 gives the annual average rates.

Next, agriculture and forestry can change carbon sinks, in particular the organic components of soil (Schoedder, 1990). Thus, the carbon household may change as a result of changes in the amount of fixation of carbon in the soil organic matter (SOM). The main sources for soil organic carbon are plant crop wastes and animal wastes. Organic matter in soil is made up of plant debris or litter in various stages of decomposition and also includes microbial biomass (see Sinabell, 1996).

During the decomposition process the microbial and chemical activities transform most of the organic mass of plants to CO_2 within a relatively short time. Only a very small part is finally transformed into persistent substances. The main determinant of the decomposition rate (other factors held constant) is the content of lignin in plants that varies strongly among different crops. Therefore crops can be classified according to their potential for humus accumulation (Haider, 1992).

Table 10.3: Annual net primary production of dry matter biomass by type of vegetation

Vegetation type in temperate climate	Net primary production of biomass		Corresponding	
	(kg/m^2)	(t/ha)	C_{org} (t/ha)	CO_2 (t/ha)
Coniferous wood	1.3	13.0	6.0	22.0
Deciduous wood	1.2	12.0	5.5	20.3
Grassland	0.8	8.0	3.7	3.2
Arable land	0.7	7.0	3.2	11.8

Source: NPP: Whittaker and Likens (1975); for the corresponding carbon amounts the conversion factors of Körner (1989) have been used.

Other factors that affect the stability of SOM are soil texture, soil structure, soil fertility, the soil's preservation capacity and the vegetation-soil effect (although in forests more debris reaches the soil surface, the decomposition is faster in forest soils than in grassland soils). In forests most of the organic matter, which is added as litter, is decomposed before it is incorporated into the soil. In grassland, additions occur by expansion of the root system, which is equally distributed throughout the soil profile (Bouwman, 1989, p.66).

As stated, the soil content of organic carbon is governed by the soil organic matter (SOM). The level and range of the SOM content is primarily determined by the type of land use. Average values of SOM in humid regions, relevant for the Alpine area, are given in Table 10.4. It is shown that changing the land use from forest to arable land is associated with a loss of up to one-third of the organic C content. Turning arable land into grassland, however, works in the opposite direction and can double organic C content and thus fix a considerable amount of atmospheric CO_2.

Table 10.4 shows that a change in land use from arable land to grassland is associated with an increase of 95 t carbon per ha bound in the soil, which is equivalent to fixing an additional 348 t CO_2 per ha. The opposite effect is true when grassland is turned into arable land where approximately half of SOM is lost within a few years, assuming no input other than crop residues is available.

SOM varies not only with land use, as seen in Table 10.4, but also with management practice. Experimental data clearly show that recycling

Table 10.4: Average content of soil organic matter (SOM) in soils in humid regions

Land-use type	SOM percent	SOM (t/ha) in topsoil	C_{org} (0–100 cm)
Deciduous wood	4	200	100
Coniferous wood	6	240	120
Grassland	7	350	175
Arable land	2	160	80

Notes: According to the literature approximately 50% of SOM is present as C_{org} (Schroeder, 1984). Sinabell (1996) further assumes that 1 m^3 of soil is equivalent to 1 500 kg at an average density.

Source: Schroeder (1984); Sinabell (1996).

farm manure on arable land can lead to a permanent increase of SOM in a given plot. For this chapter, however, we focus solely on the change in the content of organic matter (SOM) with respect to land use, as documented in Table 10.4.

While industrial and service sectors contribute to the build-up of greenhouse gases in the atmosphere mainly by burning fossil fuels, and less so by 'consuming' natural sinks (i.e. by changes in land use for housing and transport infrastructure), the reverse is true for agriculture and forestry. Here, land-use changes dominate.

Analysing in more detail, the most important vegetation type of forest land use in Alpine areas is coniferous forest. The nature of carbon matter in coniferous forest soils has been thoroughly documented in the literature. Focusing on several picea abies locations, which are typical for the Austrian alpine area, for example, the carbon matter in the soil is given at 84.2–130.1 t/ha (Bauer, 1989). This relatively small range also indicates that the balance of organic C in the soil is not or only very slightly influenced by the development stage of a forest. A comparison of carbon matter in the soil with that of the debris layer shows that the latter reaches only around 30% of the total carbon matter of the soil. An interdependency between the debris layer and the soil cannot be observed in spite of the great variability within the biomass processes. This can be explained by the fact that in general forests have already reached a state of balance. A further reason for this is the recognition of the atmosphere acting as buffer (Bauer, 1989). The carbon matter range given in this paragraph corresponds well with the average for forests in general as given in Table 10.4 (which refers to Sinabell, 1996).

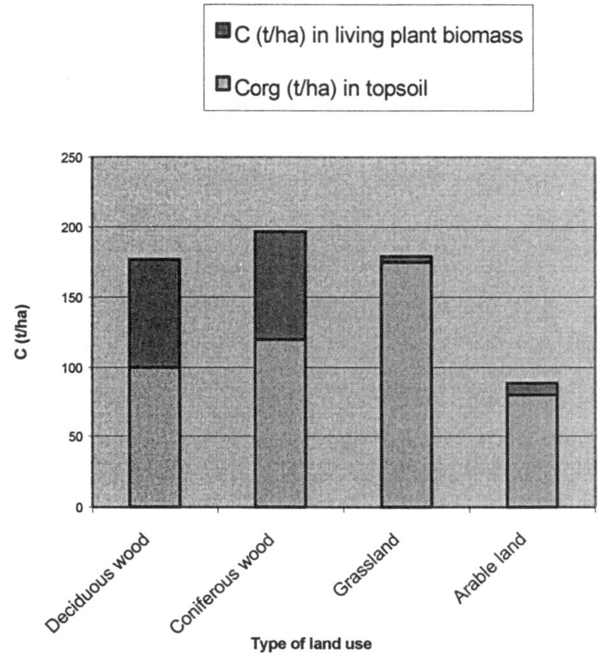

Figure 10.1: Carbon fixation by type of land use: topsoil and vegetation

The type of land use determines carbon fixation in both soil and living plant biomass. While for grassland and arable land the latter is negligible, this is not true for forest land. For land used for forestry Halbwachs et al. (1993) quantify the carbon fixed in soil at 123 t C_{org}/ha, and the carbon fixed in living plant biomass (wood) at 77 t/ha. Thus, interestingly, the sum of the two is roughly equivalent to the per hectare carbon fixation in grassland. While in the case of forests carbon fixation in soil is lower than in grassland, the difference is almost exactly matched by the carbon fixation in plant biomass (see Figure 10.1).

3.3 Production of Greenhouse Gases by Livestock

In terms of greenhouse impact non-energy related emissions of greenhouse gases are dominated by CH_4 and N_2O. CH_4 is produced mainly by livestock. The emissions of N_2O arise from three main sources:

1. usable agricultural areas;
2. forestry areas;
3. livestock.

When discussing the climate change impact of greenhouse gases the global warming potential (GWP) of these gases has to be considered (cf. Table 10.5).

Table 10.5: Lifetime and global warming potential of selected gas substances

Gas substance	Lifetime (years)	Global warming potential (100 years)
CO_2	Varies	1
CH_4	12 (+/–3)	120
N_2O	120	310
C_2HF_6	2 600	7 000
SF_6	3 200	23 900

Source: IPCC, Working Group I Report (1996).

Due to the uncertainty with respect to N_2O sources and sinks, and especially the important role of agriculture in this uncertainty, in this survey we focus only on the greenhouse gas CH_4, and thus on the source livestock.

Difficulties in quantifying N_2O sources and sinks

The chemical processes that lead to an emission of N_2O from soil are nitrification and denitrification. Both are microbacterial transformation processes of nitrogen. Under agricultural influence, however, the nitrogen turnover is higher. This is mainly caused by agricultural land management and the application of mineral and natural fertiliser. The production of N_2O depends on the concentration of NO_3^- or NO_2^-, O_3, C, acidity, temperature and enzyme state. Owing to the multiplicity of the factors involved quantification of N_2O production is extremely difficult. Long-term analyses over large areas are called for. Existing estimates range widely, and the whole topic is still relatively unknown (e.g. IPCC, 1996, 1997). For Austria overall, for example, the estimated amount of agricultural N_2O emissions is between 3.5 and 6.3 mio t N_2O–N/a (Amon et al., 2000).

Quantification of CH_4

It can be assumed that per head of cattle 81 kg of CH_4 are emitted per year. The 81 kg consist of 65 kg metabolic CH_4 and 16 kg which are due to manure storage (see Boxberger et al., 1994, p.45). This figure is used as a basis for further calculations at the farm and regional levels. The global warming potential of CH_4 is taken to be 120 times that of atmospheric CO_2 (according to Table 10.5).

4. QUANTIFICATION OF THE CURRENT CONTRIBUTION TO CLIMATE CHANGE

4.1 Approach for quantitative assessment

In order to derive the climate change impact of agriculture and forestry we draw on research results from a representative survey at the farm level. Within the research and development project 'Sustainable agricultural land use in mountainous regions of the Alps'[4] an array of indicators was defined in order to measure sustainability at both the farm and regional level. These indicators include ecological, economic and social factors. In total 43 indicators, several composed of sub-indicators, were established. Further, data to establish these indicators at the farm and regional level was collected in five research areas spread over the Alps. For this, in each participating country of the research project SAGRI-ALP (France, Germany, Italy, Switzerland and Austria), one representative research area was chosen. The selection process took account of the Alpine national farm and management characteristics. In each research area the most important farm types (according to the European Farm Accountancy Data Network (FADN) classification) with respect to land use, livestock and farm management were determined and representative farms selected. Using the regional data of the research areas and the farm data of the representative farms within these areas, the 43 different indicators were then established. In total, roughly 80 farms were surveyed within the five countries.

We draw on these results and derive the climate change contribution of agriculture and forestry from them. Where possible we do so for the Alpine area as a whole. However, as we are interested here in the relevance of agriculture and forestry for atmospheric greenhouse gas concentration, we require some reference level. We therefore decided to express changes in the carbon household arising from agriculture and forestry activity in relation to overall CO_2 emissions. The latter is best available not at any Alpine geographic definition, but at the national level, whichever particular country is chosen. For illustrative purposes we chose Austria and related the changes in carbon fixation caused by agriculture and forestry, which we derived from the Austrian research area, to Austrian annual national carbon dioxide emissions.

Let us now describe the Austrian research area in more detail. Within the 'Oberes Drautal', an east–west-oriented valley in the southern Austrian province of Carinthia, 16 farms originating from 5 farm types were chosen for the survey at farm level. The chosen research area is typical for a large part of the Austrian Alps, in particular for rural areas in the southern and central Alps, in terms of socioeconomic and land management characteristics. This is true with respect to the relation between arable land and permanent grassland, the tendency in quantitative changes of land use and livestock.

In the following section the indicators relevant for climate change are given and quantification results for the Alpine area in general, as well as more detailed results with respect to the Austrian Alpine area, are presented.

4.2 Definition of Indicators and Quantitative Results at Farm and Regional Level

Direct energy consumption

The indicator covering the consumption of direct energy includes the following types of energy: fuel (diesel oil, petrol), gas and electricity. Energy content is measured in gigajoules (GJ). In order to ensure comparability among the research areas the simpler 'end-energy' approach was chosen.

In accordance with Diepenbrock et al. (1995), specific conversion factors for the different types of energy consumption were used (see Table 10.6).

Overall Alpine results Results in direct energy consumption for the 80 farms investigated in the five research areas vary – mainly with respect to farm type - from 0.9 up to 10.2 GJ/yr. The majority of the farms investigated shows direct energy consumption of about 5 GJ/yr. There is no significant difference among the research areas.

Austrian results Using the survey data at farm level and the conversion factors of Table 10.6, the average farm in the Austrian survey area consumes 5.57 GJ/ha/yr in diesel consumption and 1.77 GJ/ha/yr in electricity consumption. Using the CO_2 content of these energy types (0.075 kg CO_2 per GJ diesel oil and 0.058 kg CO_2 per GJ electricity) agriculture in the Austrian research area 'Oberes Drautal' produces 10.21 t CO_2 per year.

If we extrapolate the average farm results in the research area to the whole Austrian Alpine area we get a direct energy consumption related emission of 406.7 t CO_2 per year. This is less than 0.001% of total Austrian annual CO_2 emissions, which is thus also a remarkably low portion of the overall agricultural 7% share in CO_2 emissions (including the non-Alpine areas). It should therefore be obvious that CO_2 production in Alpine agriculture as a result of direct energy consumption is of almost no importance on a national scale. That the area of usable agricultural land in the Austrian Alps has been steadily decreasing in recent years (–4.9% in the time period 1983–1995) only lends further support on this conclusion.

Table 10.6: Conversion factors for direct energy consumption (energy input following the end-energy approach)

Type of consumption	Unit	Content energy-input
Fuel (diesel oil)	l (litre)	0.0358 GJ
Fuel (petrol)	l (litre)	0.0313 GJ
Gas	m3 (cubic metre)	0.0350 GJ
Electricity	kWh (kilowatt-hour)	0.0036 GJ

Source: Diepenbrock et al. (1995, pp.14–17).

Indirect energy consumption

Indirect energy consumption covers the application of mineral fertiliser, agricultural pesticides and bought feed.

Overall Alpine results The survey at farm level in the different research areas shows that Alpine agriculture in general is based on rather extensive production techniques; the consumption of fertiliser and pesticides is rather small. Obviously this depends on the orientation of the specific farm. Farms with more of a permanent grassland orientation (the majority of the Alpine farms) consume only small amounts of these substances. Only farms with arable land and permanent cultures such as vineyards or orchards lead to noticeable consumption of such substances. Considering the share of these types of land use in the total Alpine farmed area (at the European level 5.2% for arable land and 0.8% for permanent cultures), these types of land use are clearly of low importance.

Austrian results In Austria these types of land use are in fact below the overall Alpine average; there is only 3.7% arable land and 0.1% permanent culture (see Table 10.2). The CO_2 production caused by indirect energy consumption can thus be considered negligible. This is in strong contrast to agriculture in non-Alpine areas (for example, expenditure on fertiliser is only 0.5% of total farm expenditures in Alpine areas, but between 2 and 4% in non-Alpine areas (BMLFUW, 2000: 241)).

Carbon pool in Alpine agricultural and forest soils

On applying the SOM data by soil type presented in section 3, the carbon pool in agricultural and forestry soils was calculated for the Alpine area. The following simplifications apply. As no distinction in overall land-use data is made between deciduous wood and coniferous wood, coniferous wood is considered representative for total forest. This is done both because the coniferous wood dominates within the Alpine area, and because the respective content of C_{org} of coniferous and deciduous wood is rather similar. Permanent cultures are treated as arable land as no separate SOM figures were available.

Table 10.7: Greenhouse relevant emissions and fixation in the Alpine area

	Total Alpine area	Austrian Alpine area	Evaluation
Direct energy consumption		406 t CO_2/year	Negligible
Indirect energy consumption (mineral fertiliser, pesticides, bought feed)	Negligible	Negligible	Negligible in Alpine area
Carbon pool in soil	3.4 bill. t CO_2	1.5 bill. t CO_2	
Production of CH_4 by livestock	GWP: 32.4 m. t CO_2/year	GWP: 11.4 m. t CO_2/year	

Overall Alpine results Proceeding in this way, we find that the *categories Agricultural usable area* and *Forestry usable area* represent an amount of 930 million t C in topsoil matter. This means that an equivalent of 3 410 million t CO_2 is fixed by the soil (see Table 10.7).

Austrian results With respect to the research area 'Oberes Drautal' 5.12 million t C_{org} are in soil matter, giving fixation of about 18.8 million t CO_2. For the total Austrian Alpine area the amount of C_{org} in the soil matter is about 415 million t. This represents a fixation of roughly 1.5 billion t CO_2.

Production of CH_4 by livestock

Applying the amount of CH_4 production per head of cattle, as given in section 3, the amount of production of CH_4 by the total livestock within the Alpine area can be calculated.

Overall Alpine results Based on agricultural statistics, currently cattle livestock within the Alpine area amounts to approximately 1 177 000. The Alpine cattle livestock thus produces an amount of 270 300 t CH_4 per year. The global warming potential of this is equivalent to 32.4 million t CO_2.

Austrian results Within the Austrian research area, CH_4 emission is around 707 t/yr which represents a global warming potential (GWP) of the equivalent of 84 840 t CO_2 per year. For the total Alpine area of Austria this implies emissions of about 95 340 t CH_4 per annum which corresponds to 11.4 million GWP.

5 SCENARIOS FOR AUSTRIA

5.1 Recent Developments in Changing Land Use

Within the Austrian research area, a slight increase of forest and a slight decrease of grassland and arable land were observed within the time period of 1983–1995. On a yearly average this trend has reduced CO_2 fixation by 1.6 t/yr. The slight decrease of livestock implies an average yearly decrease of CH_4 emissions of about 2 t/yr.

For the total Alpine area of Austria an increase of both forest and grassland and a decrease of arable land took place. Additionally there was an increase in livestock. Transformed into greenhouse gas emissions, yearly CH_4 emissions increased at an average rate of 12 t/yr, but the CO_2 fixation in the soil matter increased by 1.25 mio t/yr.[5] The latter impact of carbon fixation clearly dominates the overall greenhouse impact of this trend.

5.2 Description of Three Realistic Future Scenarios

We employ three scenarios to analyse possible future development. These were defined within the research project SAGRI-ALP cited earlier. Each of these scenarios describes possible changes in land use and livestock, which we use here to calculate respective impacts on the climate-relevant indicators.

The basic orientation of each of the three scenarios is as follows:

Scenario 'trend' This scenario analyses the impact of the Agenda 2000 policy, including treatment of less favoured areas (Reg. 950/92), agri-environmental measures (Reg. 2078/92), premia for livestock breeding and further measures according to the rural development plan. In accordance with the Austrian programme for agri-environmental measures ('ÖPUL 2000', based on Council Regulation (EC) 1257/1999 and Commission Regulation (EC) 1750/1999), and in accordance with the incentives for the less-favoured areas the premia for livestock are taken to increase.[6]

Scenario 'local people' This scenario considers the point of view of local people, derived by means of several discussions within local working groups. The members of these local working groups were mainly farmers, but politicians and representatives from other sectors were also involved. The main orientation of this scenario is the following. First, a reduction of the workload for the farmers and their families is assumed to take place and second, financial support for foreign work forces (temporary and permanent labour) is assumed.

Scenario 'scientists' This represents the point of view of the scientific team of SAGRI-ALP. In general, ecologically oriented aspects are considered. The overriding aims of this scenario are an implementation of agri-environmental incentives at the highest possible level, and prevention of abandonment of rural areas. This leads to more extensive land use where the upper limit for arable land area is set at 20% of total farm land and no more than two crops are allowed for grassland, and restrictions on the planting of silage maize are imposed.

5.3 Results of the Three Scenarios with Respect to the Climate-relevant Factors

Scenario 'trend'

The higher premia for livestock as proposed by the Austrian programme for agri-environmental measurements (known under the acronym ÖPUL 2000) do not have an impact on production within the surveyed farms. Basically, marginal revenues of individual activities are changed, but not enough to switch the break-even point of the highest revenue activity to a different activity. Thus, there is only an economic effect of a slight increase of the gross margin – especially for the bigger farms with a clear cattle orientation, but climate-relevant factors (land use, number of livestock) remain unchanged.

Table 10.8: Climate-relevant factors of the scenario 'trend' listed for the surveyed farms of the Austrian research area

Farm type	Land use (ha)		Livestock unit (LU)	
	actual	future	actual	future
(1) Specialised suckler cow farms	11.3	11.3	11.9	11.9
(2) Specialised dairy farms	18.8	18.8	39.7	40.0
(3) Specialised cattle rearing and fattening	19.3	19.3	70.0	70.0
(4) Specialised forestry farms	24.9	24.9	70.0	70.0
(5) Specialised farms with grazing livestock	7.1	7.1	6.0	7.5

Source: SAGRI-ALP (2000) – Regional Report 3, Research Area 'Oberes Drautal'.

Table 10.8 indicates that the total amount of managed land remains unchanged. Also the relation between arable land and permanent grassland (not stated in Table 10.8) remains stable. Slight changes occur only in the number of livestock, and then only in two of the five farm types analysed. The specialised dairy farms increase livestock from 39.7 to 40 livestock units (LU) and the specialised farms with grazing stock (sheep and goats) from 6 to 7.5 LU. In Table 10.8 the change in the type of livestock is not visible. The specialised dairy farms and the specialised farms with grazing livestock (sheep and goats) exchange a part of their heifers for cows. This is induced by the higher premium received for cows. But this change has no effect on the total number nor on livestock unit, and therefore no influence on the climate-relevant factors. In general almost no change in the climate-relevant factors occurs.

Scenario 'local people'

Even when those aspects believed to be most important by local people are implemented, no change in land use occurs. All types of farms show a constant amount of managed land. Thus, neither has additional land been purchased or rented for management nor has currently managed area fallen out of management. The relation between arable land and permanent grassland also remains constant.

Additionally, in this scenario no change in the number of livestock can be observed. The only changes that take place are among the type of livestock. The specialised dairy farms substitute heifers for dairy cows due to the lower effort needed in their upkeep. With a decreasing price for foreign labour, heifers are replaced by dairy cows. Whatever the case, such substitution effects are not relevant in the present analysis since in total, there is no noticeable effect on climate-relevant factors.

Scenario 'scientists'

In the scenario 'scientists' several changes occur, as is evident from Table 10.9. In this scenario the total sum of managed areas again remains constant, and is therefore not stated in Table 10.9. This is a direct result of the scenario objectives, the most important being the prevention of abandonment of land. However, within land use a strong decrease of arable land combined with an increase of permanent grassland can be observed.

Table 10.9: Climate-relevant factors of the scenario 'scientists' listed for the surveyed farm types of the Austrian research area

Farm type	Grassland (ha)		Arable land (ha)		Livestock unit (LU)	
	actual	future	actual	future	actual	future
(1) Suckler cow	11.3	11.3	0.0	0.0	11.9	8.5
(2) Dairy	10.7	15.0	8.1	3.8	39.7	24.8
(3) Cattle	10.3	15.4	9.0	3.9	70.0	32.0
(4) Forestry	21.1	21.1	3.8	3.8	70.0	39.9
(5) Grazing	7.0	7.0	0.1	0.1	7.5	6.0

Source: SAGRI-ALP (2000) – Regional Report 3, Research Area 'Oberes Drautal'.

As Table 10.9 indicates, farm types 2 and 3 (specialised dairy farms, specialised cattle-rearing and fattening farms) massively reduce their arable land under operation. This leads to an average reduction of arable land by 45% over the scenario time period of seven years.

There are also dramatic changes in livestock. As Table 10.9 indicates, in all farm types the number of livestock units decreases, by between –54% (specialised cattle-rearing and fattening farms) and –20% (specialised farms with grazing livestock – sheep and goats). The average decrease in livestock in the Austrian research area is around 37%.

5.4 Quantification of Climate Relevant Impacts

With respect to climate–relevance it is evident that both the *'trend scenario'* and the *'local people scenario'* do not indicate any relevant aspect that leads to a change in the contribution of agriculture and forestry to climate change. In both scenarios the agro-environmental incentives are either too small or the effects that do occur tend to neutralise each other.

Only the *'scientists' scenario'* leads to a noticeable effect, mainly due to land-use change, but also due to livestock-changes. Applying the data given in sections 3 and 4 to the decrease of arable land in this scenario, and assuming that all of that arable land is transformed into grassland or forest (as given in Table 10.9), which has equivalent implications in terms of carbon fixation, the following change in the carbon balance can be determined. For the specific research area a sum of 843 ha arable land is transformed into grassland in this scenario (multiplying the land use-change of the average farm as given in Table 10.9 with the number of farms in each farm type) and for the total Austrian Alpine area a total of 91 035 ha is transformed into grassland or forest accordingly. This represents an additional fixation of around 293 000 t CO_2 within the research area, and about 31.6 million t CO_2 for the total Austrian Alpine area. Livestock units are reduced in the research area in this scenario, implying a reduction in CH_4 emissions of 212 t per annum. Extrapolating for the total Austrian Alpine area this reduction amounts to a GWP of 3.4 million.

5.5 Quantification of (Unrealistic) Extreme Scenarios and their Climate Relevant Impacts

Just to get an impression of the overall maximum quantitative importance of Alpine agriculture in the carbon household, the following hypothetical experiment can be done. We can imagine a maximum *'grassland scenario'* for Alpine Austria, characterised by a total transformation of all arable land within the Austrian alpine area into grassland. Or taking the opposite case, we can imagine a maximum *'arable land scenario'* whereby all grassland is transformed into arable land. The consequences for the carbon household can then be calculated. Using the factors described in section 3 (based on Table 10.4) the long-term equilibrium of the grassland scenario points to a fixation of an additional 70.4 million t CO_2 equivalent. The other extreme, 'arable land scenario', would result in a loss of a sink for 197.9 million t CO_2. While such extreme changes remain hypothetical, the move from arable land to grassland has recently been observed in Austria (see section 5.1).

6 CONCLUSIONS

This chapter explored the quantitative potential of Alpine agriculture and forestry for climate change mitigation. The most relevant factors of climate-relevance were identified as (a) changes in the carbon household due to land-use changes and (b) production of greenhouse gases by the livestock. The quantitative relevance of these factors together with greenhouse gas emissions from direct and indirect energy consumption was then given on the basis of a farm-level survey in representative Alpine research areas in five countries.

Further, for Austria, three future scenarios from a recent study on Alpine agricultural and forestry development were analysed with respect to their implications for the greenhouse gas balance. We found that neither the trend scenario nor the scenario taking account of the objectives of the local people induce any changes in land-use or livestock that would be of a relevant magnitude for the overall greenhouse balance. From the three scenarios presented here, only that reflecting the opinions of the scientific research team undertaking the European Union-funded research project on sustainable agricultural land-use in mountainous regions of the Alps (SAGRI-ALP) provided any significant indication of changes in carbon fixation as a result of alterations in land use patterns and changes in livestock greenhouse emissions. The two major effects found were the following. The first, arising from a shift away from arable land, involves a one-off additional fixation of carbon, which is in fact equivalent to approximately half of Austrian annual CO_2 emissions. The second effect, arising from the predicted fall in livestock, is equivalent to a reduction of Austrian CO_2 emissions by 5.7% in terms of GWP.

Finally, the most extreme (and unrealistic) scenarios of a total transformation of all Alpine arable land into grassland in Austria or reversal, all grassland into arable land, are quantified with respect to their greenhouse gas balance implications. In the first case, an additional fixation of carbon would occur, roughly equivalent to Austria's current CO_2 emissions in a single year, while in the second case a loss of a carbon sink binding the amount of three years of current Austrian CO_2 emissions would occur. These calculations are meant to indicate the maximum possible impact that Alpine agriculture and forestry could have in climate change mitigation. However, the earlier stated scenario (scientists scenario) implies transforming roughly half of the Austrian Alpine arable land into grassland or forest and has been considered a potentially realistic scenario. This would fix half of a single year's CO_2 emissions in Austria.

NOTES

1. The definition of the Alpine Convention covers roughly 5,600 communities in Germany, France, Italy, Liechtenstein, Austria, Switzerland and Slovenia. Compared to the definition of the Alpine area by Bätzing (1993) the Alpine Convention covers a slightly larger area. This is true in particular in the setting of the northern border of the Alps.
2. Project-Acronym: SUSTALP; Project N0. ENV4- CT97-0442; SUSTALP, 2000. The Project partners of the SUSTALP research project are the European Academy (Bolzano, Italy; project coordinator), the Alpenforschungsinstitut GmbH (Garmisch – Partenkirchen, Germany), the RaumUmwelt PlanungsGesmbH (Vienna, Austria), the Österreichische Vereinigung für Agrarwissenschaftliche Forschung (Vienna, Austria) and the University of Augsburg (Augsburg, Germany).
3. The classification of the sectors is chosen according to the EU classification NACE. Within the so-called primary sector the sectors of agriculture, forestry and fishery are included.
4. R&D project of the European Commission, DG VI; 'Sustainable agricultural land use in alpine areas' (SAGRI-ALP; FAIR5 CT97-3798)
5. The Alpine arable land in Austria declined by 43 000 ha in the period 1983 to 1995. It is assumed that all of this land was transformed either into grassland or forest use, which in terms of total carbon fixation (soil plus biomass) is roughly of equal impact (see Figure 10.1).
6. The premia structure is taken to change as follows: suckler cows: +21% if the livestock units (LU) per ha are above 1.4 or +77% if the livestock units per ha are below 1.4; fattening cattle: +12.4% if LU per ha is above 1.4 or +58.5% if LU per ha is below; sheep: +2%.

BIBLIOGRAPHY

Amon, B., Th. Amon and J. Boxberger (2000), 'N_2O aus Land- und Forstwirtschaft', in A. Hackl and S. Schleicher (eds.), *Lachgasemissionen und Klimaschutz*, Austrian Council on Climate Change and Bundesministerium für Umwelt, Jugend und Familie, Wien: Service Fachverlag, p.1–11.

Bätzing, Werner (1993), *Der sozioökonomische Strukturwandel des Alpenraumes im 20. Jahrhundert. Eine Analyse von "Entwicklungstypen" auf Gemeindeebene im Kontext der europäischen Tertiarisierung*, Bern, Geographica Bernesia.

Bauer, Hubert (1989), '*Nährstoffvorräte von Fichtenbeständen auf einer Standortseinheit im Kobernauserwald untersucht über die Altersklassen'*, Diplomarbeit durchgeführt am Institut für Forstökologie an der Universität für Bodenkultur, Wien.

BMELF (1983), 'Energieprobleme und die Nahrungswirtschaft', in *Schriftenreihe des Bundesministeriums für Ernährung, Landwirtschaft und Forsten*, Reihe C, Heft 16, Münster Hiltrup, Landwirtschaftsverlag.

BMLFUW (2000), *Grüner Bericht 1999*, Wien: Bundesministerium für Landwirtschaft, Forstwirtschaft, Umwelt und Wasserwirtschaft.

Bouwman (1989), 'Exchange of Greenhouse Gases Between Terrestrial Ecosystems and the Athmosphere', in A.F. Bouwman (ed.), *Soils and the Greenhouse Effect*, Chichester, John Wiley and Sons, p.75–97.

Boxberger, J., Th. Amon and H. Dissemond (1994), 'Biogene Emissionen der Landwirtschaft', in Bundesministerium für Umwelt, Jugend und Familie and Österreichische CO_2- Kommission (ACC) (ed.), *CH_4-Emissionen in Österreich* 37-50. Reihe: Dokumentation, Bd. 6. Laxenburg. Akademie für Umwelt und Energie.

Diepenrock, Wulf, A. Pelzer and Jörn Radtke (1995), 'Energiebilanz im Ackerbaubetrieb' in KTBL (Kuratorium für Technik und Bauwesen in der Landwirtschaft e. V., ed.) KTBL-Arbeitspapier 211, Münster-Hiltrup, Landwirtschaftsverlag.

Haider, K. (1992), 'Biochemische Prozesse der Bildung und der Dynamik von Huminstoffen im Boden', in Bundesministerium für Ernährung Landwirtschaft und Forsten (ed.), Bodennutzung and Bodenfruchtbarkeit. Band 4 Humushaushalt 45–62. Reihe: *Berichte über Landwirtschaft*. Neue Folge. 206. Sonderheft. Hamburg, Verlag Paul Parey.

Halbwachs, G., N. Arzl, M. Grabner and R. Wimmer (1993), 'Kohlenstoffflüsse in der österreichischen Forst- und Holzwirtschaft II', in Akademie für Umwelt und Energie, Bundesministerium für

Umwelt, Jugend und Familie und Österreichische CO_2-Kommission (ACC) (eds.), *Jahresbericht 1992, Empfehlungen 1992 der österreichischen CO2 Kommission für ein Aktionsprogramm zur Erreichung des Toronto-Zieles*, Reihe Forschung, Bd. 3, Wien.

IPCC (1996), *Climate Change 1995 – Impacts, Adaptions and Mitigation of Climate Change: Scientific – Technical Analyses.* Contribution of Working Group I to the Second Assessment Report of the Intergovernmental Panel on Climate Change, R. T. Watson, M. C. Zinoyowera, · and R. H. Moss (eds.), Cambridge/New York, Cambridge University Press.

IPCC (1997), *Guidelines for National Greenhouse Gas Inventories*, Paris, OECD/OCDE.

Körner, Christian (1989), 'Bedeutung der Wälder im Naturhaushalt einer vom Menschen veränderten Welt', in Österreichische Akademie der Wissenschaften (ed.), *Veröffentlichungen der Kommission für Humanökologie 1, Die Bedrohung der Wälder – Schäden, Folgeerscheinungen und Gegenmaßnahmen*, Wien, Verlag der Österreichischen Akademie der Wissenschaften, pp. 7–40.

Österreichischer Klimabeirat (ACCC, 2000), Die österreichische CO_2-Bilanz 1998 – Struktur und Dynamik der österreichischen CO_2-Emissionen, Austrian Council on Climate Change (ACCC, ed.), Internet www.accc.gv.at.

Reinhardt, G.A. (1993), *Energie- und CO_2–Bilanzierung Nachwachsender Rohstoffe*, Braunschweig/Wiesbaden, Vieweg.

SAGRI-ALP (2000), Regional Report 3 of the Austrian research area 'Oberes Drautal', Regional results of task 4 within the research project 'Sustainable Agricultural Land use in Alpine Mountain Regions (SAGRI-ALP)', unpublished manuscript.

Schoedder, F. (1990), 'Entstehung klimarelevanter Spurengase als Folge der Landbewirtschaftung' in Landbauforschung Völkenrode Wissenschaftliche Mitteilungen der Bundesforschungsanstalt für Landwirtschaft Braunschweig – Völkenrode (FAL), Sauerbeck, D., H. Hummert (eds.), *Studie der ad-hoc-Arbeitsgruppe Klimaveränderung und Landwirtschaft des Senates der FAL*, part 1, Braunschweig, Landbauforschung Völkenrode, pp.17–27.

Schroeder, D. (1984), *Bodenkunde in Stichworten*, Unterägeri, Hirt.

Sinabell, Franz (1996), 'Soil as a sink of atmospheric CO_2 – the potential contribution of Austrian agriculture', in Österreichische Gesellschaft für Agrarökonomie (eds), *Proceedings of the annual meeting of the Österreichische Gesellschaft für Agrarökonomie*, Kiel, Wissenschafts-Verlag Vauk. pp.79–96.

SUSTALP (2000), Results of surveys of national agrarian statistics within the research project 'Evaluation of Instruments of the European Union regarding their contribution to Sustainable Environment and agriculture in the Alps (SUSTALP)', unpublished manuscript (edited volume on the project results in preparation).

Werschnitzky, U., G. Prothmann and C. Sergeois (1987), 'Direkter Energieeinsatz im agrarischen Erzeugerbereich in der BRD' in Kuroatorium für Technik und Bauswesen in der Landwirtschaft (KTBL) (ed.), *Energie und Agrarwirtschaft*, KTBL-Schrift 320, Landwirtschaftsverlag, Münster-Hiltrup, pp.13–49.

Whittaker, R.H. and G.E. Likens (1975), 'The biosphere and man' in H. Lieth and R.H. Wittaker (eds.), *Primary Productivity of the Biosphere*, Ecological Studies 14, Berlin, Springer, pp.305–328.

11. Climate Change Mitigation in the Alps by Means of Renewable Energy Use: The Austrian Province of Carinthia as an Illustration

Reinhard Madlener, Anna Kleissner and Norbert Wohlgemuth

1 INTRODUCTION

Irrespective of the region studied, the increased use of renewable energy as a substitute for fossil fuels has been widely recognised as one important strategy in combating the anthropogenic greenhouse effect.[1] Moreover, locally available renewable energies often provide ancillary socioeconomic-ecological (SEE) net benefits that offer a chance for a more sustainable development – like employment and value-added creation in rural and/or remote areas, balance of trade improvements, or increases in energy self-sufficiency, supply security and supply diversity. Such ancillary benefits can help to sustain rural communities and local enterprises engaged in the production of (ideally state-of-the-art) renewable energy technologies, or of biomass fuels.

An important aspect in discussions on the diffusion of modern renewable energy technologies is their cost competitiveness (or rather the lack of it in many of the typically simplistic comparisons). Fortunately, in recent years and spurred by targeted policy efforts, numerous technological improvements have raised the attractiveness of many renewable energy technologies considerably in terms of their cost effectiveness, reliability, availability, handling and environmental impact. For this reason the range of feasible and desirable renewable energy options for a specific location ought to be reconsidered from time to time. Commonly, little is known at the regional level about the achievable and cost-efficient contribution of renewable energy use to climate protection (as well as to other SEE aspects). On the one hand, renewables compete with possibly more cost-efficient measures for mitigating greenhouse gas (GHG) emissions, such as improvements in energy efficiency on the demand side (e.g., better housing insulation) or the supply side (e.g., combined heat-and-power production). On the other hand, they also compete with each other and with fossil- and nuclear-based supply alternatives. The problem, especially with the latter kind of competition, is the fact that in most cases a truly level playing field does not exist, e.g., due to subsidies granted to conventional energy technologies, or because external costs are not sufficiently accounted for, which leads to market failure.

Natural endowments with renewable energy resources, societal preferences and the existence of local renewable energy technology manufacturing industries and energy service providers can all be expected to vary from region to region. Generalisations are, therefore, rather difficult. This is especially true for mountain regions like the European Alps, where the use of solar, hydro, biomass and wind energy resources is often particularly attractive, and where – due to the higher ecological and social sensitivity of the area – frictions between the mountain and the lowland (and often densely populated) areas, can be felt particularly strongly. Frictions can also be expected between globally oriented policies (comparative advantages, economies of scale) and locally oriented policies (self-sufficiency, local sustainability). At the same time, the mountains clearly provide the economic, social and cultural realm for all their inhabitants (e.g., Price et al. 2001).

Until the first oil shock in the 1970s, renewables were widely considered to be a minor source of energy and of declining importance. Countries like Austria and Switzerland with large shares of hydro-power schemes in overall electricity production were perhaps exceptions in this respect (1999: Austria: 69%; Switzerland: 61%). In the meantime, environmental, (long-term) economic, and to an increasing extent the social benefits of renewables use have become a major driving force for their increased implementation. Within the European Union, for example, there are numerous policy initiatives in place at different levels to promote the use of renewable energies in line with the Commission's White Paper on renewable sources of energy (EC 1997b) and the new EU Directive on the promotion of electricity from renewables (EC 2001b).

Having said all this, however, it should be also noted that renewable energy use in many applications can have non-negligible (and often very site-specific) adverse impacts as well. Systematic and extensive discussions of these have only more recently found their way into the literature (e.g., IEA 1998; Dincer and Rosen 1999; Abbasi and Abbasi 2000), probably triggered by the rapidly increasing levels of some renewable energy technologies (e.g., wind power) and the generally revived interest in the use of renewables as a result of the climate change and sustainability debate.

With the increasing liberalisation and harmonisation of the European internal electricity market (EC 1997a) and gas market (EC 1998),[2] short-term cost competitiveness has been gaining in importance, leading to potential conflicts with long-term economic, social and ecological goals (e.g., climate protection, biodiversity, beauty of the landscape). Such conflicts are at least partly due to the high degrees of uncertainty and subjectivity involved, which all too often make it much harder to quantify and integrate environmental impacts (for a recent discussion of integrating environmental issues and competition-oriented economic policy at the EU level see, for example, EC 2000). Consequently, for the sustainability-oriented policy-maker it is of paramount importance to evaluate the renewable energy technologies fostered in a region in a holistic way – on the basis of their overall impacts on society, the environment and the economy (thus taking account of the various dimensions of sustainable development). Also, it is clear that GHG mitigation is just one of many goals on the road to sustainability.

One way to consider the multidimensional benefits (and disbenefits) that can be reaped from renewables is the use of weights in renewable energy promotion schemes. Such weights complement the short-run cost aspect and aim at reflecting the particular impacts, e.g., in terms of GHG mitigation potential, employment creation, avoidance of local air, soil, and/or water pollution (objective criteria). Additionally, they may also reflect some consensus regarding the social preferences within the region studied (subjective criteria). Such multidimensional evaluations can be used to differentiate between the various impacts of renewable energy technologies, and, once established, they can then be used straightforwardly, for example, in 'truly' sustainability-oriented renewable energy promotion schemes. A disadvantage of the approach just described is that it can be expected to be rather cumbersome to undertake in practice.

In our empirical illustration we will mainly study the GHG mitigation that can be achieved in Alpine regions by means of the typically, locally available renewable energy sources. This will be exemplified by an analysis of Austria's most southern province, Carinthia, that draws on results from an empirical study undertaken in 1997 within the EU programme ALTENER (Verbundplan/ESD 1997). Many of the results gained from this investigation can, to a certain extent, be generalised for other Alpine regions as well, and may additionally be useful for other sensitive regions both in the developed and the developing world. Note, however, that energy-related aspects of the transportation sector, though of paramount importance to the Alpine regions because of their vulnerability and position at the crossroads of Europe, are well beyond the scope of this rather introductory chapter (for a discussion of the global environmental impacts of Alpine transport see Chapter 12).

The remainder of the chapter is organised as follows: Section 2 discusses some energy-related policy initiatives and policy aspects in the context of the Alps, section 3 deals with the impacts of globalisation and liberalisation on Alpine regions, section 4 addresses the greenhouse gas mitigation potential of renewables, section 5 covers resource, technology and social acceptance aspects and section 6 contains the case study on Carinthia. Section 7 provides conclusions.

2 POLICY INITIATIVES AND ASPECTS

In this section we want to first introduce the energy-related parts of the three political initiatives Agenda 21, CIPRA and the Alpine Convention, and then discuss some other policy-related aspects of renewable energy use specific to Alpine regions.

Agenda 21

Within Agenda 21 of the United Nations, energy issues are addressed primarily in the context of Chapter 9 'Protection of the Atmosphere'. This chapter has the objective 'to increase the contribution of environmentally sound and cost-effective energy systems, particularly new and renewable ones, through less polluting and more efficient energy production, transmission, distribution and use' (UNCED 1993).

In 1998, the General Assembly of the United Nations proclaimed the Year 2002 as the 'International Year of Mountains', and sustainable mountain development forms the subject of Chapter 13 of Agenda 21.[3] This Agenda 21 chapter on 'Managing Fragile Ecosystems – Sustainable Mountain Development' identifies mountains as 'fragile or vulnerable ecosystems', as they are characterised by close and continuous interactions between human beings and nature, between natural processes and human activities. Exactly this interdependency of natural and socioeconomic processes makes a close collaboration between social and natural scientists in studying Alpine regions indispensable. The fragile balance between these two aspects has to be studied and taken into account with regard to possible political strategies and to the needs for inter- and trans-disciplinary efforts in both the natural and the social sciences. In what follows we include a short excerpt from Chapter 13.1 for further illustration:

> Mountains are an important source of water, energy and biological diversity. Furthermore, they are a source of such key resources as minerals, forest products and agricultural products and of recreation. As a major ecosystem representing the complex and interrelated ecology of our planet, mountain environments are essential to the survival of the global ecosystem. Mountain ecosystems are, however, rapidly changing. They are susceptible to accelerated soil erosion, landslides and rapid loss of habitat and genetic diversity. On the human side, there is widespread poverty among mountain inhabitants and loss of indigenous knowledge. As a result, most global mountain areas are experiencing environmental degradation. Hence, the proper management of mountain resources and socio-economic development of the people deserves immediate action (UNCED 1993).

An estimated 10% of the world's population depends directly on mountain resources, and a much larger percentage draws on them in one form or another (UNCED 1993). A highly sensitive ecosystem, such as the Alps, responds particularly swiftly to climatic changes. Glaciers in Austria, for example, have shrunk by approximately one-third of their surface since their maximum expansion in the mid 19th century. If, as expected, the average temperatures rise by 3.8 °C, the glaciers in Austria could disappear within some 50 years and the timber line might rise by about 700 metres. As a consequence, even a heavily intensified use of artificial snow, and disregarding for the moment its potentially severe adverse environmental impact, may in many tourism-dependent regions fail to maintain winter tourism at today's levels (Hein 1996; see also Chapter 4, section 6 in this book).

The Alpine Convention and the International Alps Protection Commission (CIPRA)

According to the geographical demarcation of the Alpine region shared by the signatory parties to the so-called *'Alpenkonvention'* (Alpine Convention; see below), depicted in Figure 11.3, the Alps are inhabited by some 11.2 million people, living in eight countries in some 5 800 communities, that are spread over some 200 000 square kilometres.

The International Alps Protection Commission (CIPRA) is a body founded in 1952 that aims at improving trans-boundary exchange of information and experience of relevance to the Alps. Sustainable use of energy in the Alpine regions has become increasingly important to

CIPRA, which has observer status regarding the Alpine Convention, and a number of declarations and other material has been made available by the organisation that arranges international annual meetings on specific topics related to the Alps (CIPRA 1995, 1998; see also www.cipra.org).

The Alpine Convention was signed in November 1991 in Salzburg, Austria, by six nations of the Alpine region (Austria, France, Germany, Italy, Liechtenstein and Switzerland) and the European Union. Former Yugoslavia, which prior to the war was supposed to sign, was replaced by the Republic of Slovenia in 1993, and the Principality of Monaco joined in 1994 via a Supplement to the Protocol. The Alpine Convention came into force on 6 March 1995, and became legally binding, after the required minimum of three (out of eight) signatory countries' national parliaments had completed the ratification process. The Convention includes minimum standards for environmental policy and a commitment for the implementation of measures for the protection of the Alps and of those leading to sustainable development in everyday life. Some parts of the Convention deal specifically with energy matters in the Alpine regions (as specified in the Appendix to the Convention), such as lit. e and k of Article 2(2), which call for

- production, distribution and use of energy, which are environmentally sound, nature- and landscape-friendly;
- promotion of energy-saving measures;
- equal consideration of the interests of the local people and the interest in preserving the environment in the case of hydro-power use.

(translation R.M./N.W.)

More importantly, after years of negotiations and a rather sluggish development, an Energy Protocol (*'Protokoll Energie'*) to the Alps Convention was eventually signed by two Ministers of the Environment (Germany, Slovenia) in October 1998 in the course of the 5th Conference of the Alps in Bled, Slovenia. France followed to sign this Protocol in December 1998, Austria and Switzerland in October 2000 and Italy in February 2001. Article 1 (Objective) of the Energy Protocol states that:

> . . . the parties to the Protocol are obliged to create frame conditions and to take up specific measures in the areas of energy saving, energy production, energy transportation, energy supply, and energy use, in order to create the energy-related preconditions necessary for a sustainable development compatible with the load limits typical for the Alpine region (translation R.M.).

Moreover, the Energy Protocol contains stipulations for a reduction of the energy-related environmental pressures when optimising energy service provision for final consumers. If possible this should be done by reducing energy demand through the use of more energy-efficient technologies, the increased use of renewables for meeting the remaining demand, and an optimisation of existing non-renewable energy supply technologies. Some critics have pointed out that due to the strong pro-nuclear interests of some of the signatory countries (particularly France and Germany), the Energy Protocol lacks a clear anti-nuclear commitment. Further, the Energy Protocol (like most of the other eight existing protocols[4]) has not yet entered into force due to the slow ratification processes in the signatory countries. This procedure, however, is expected to speed up considerably in the 'International Year of Mountains' 2002.

Other Policy-related Aspects

In ecologically sensitive, intensively cultivated and (at least partially) densely populated areas of great natural beauty and richness, the way energy is provided and used is of special interest. It is particularly important in those Alpine regions which are relatively easily accessible and/or where intensive tourism plays an important role. In such a context, interdisciplinary approaches ought to be used that properly take the multidimensional role of Alpine regions into account (sensitive and fragile living area both for humans and animals, tourism, recreation, water reservoir, economic activity etc.). As any policy chosen in these areas affects various elements of life and hence has to be designed very carefully in order to maximise synergies and minimise trade-offs, this is also true for a climate change mitigation strategy that incorporates an increased use of renewable energy (Hein 1996; Flavin and Dunn 1998).

Apart from their potential to reduce GHG emissions, and as already briefly addressed in the introduction, the use of renewable energy sources can significantly contribute to many other aspects of sustainable development apart from GHG mitigation. Among them are improved air and water quality, enhanced soil preservation, flood protection, electrification of rural and remote areas, and increased employment opportunities (Stanford 1997). Moreover, the increased utilisation of renewables may lead to a more diversified resource base, reduced fuel supply and fuel price risk, development promotion through the provision of additional infrastructure, higher economic flexibility and a lower risk exposure through modular and/or small-sized energy technologies, the creation of more choice for consumers and possibly even a better overall system reliability (Wohlgemuth and Missfeldt 2000, among others).

It should also be noted that due to the large potential and importance of biomass use in Alpine regions, and in the light of changing EU and national agriculture and forestry policies, an assessment of the climate change mitigation impacts of renewable energy use can only be undertaken by taking into account the complex interrelatedness of land use and socioeconomic and environmental priorities (cf. Chapter 10 in this book).

A method that explicitly accounts for the multidimensionality of the impacts of renewable energy use and possible differences in the preferences of the different stakeholders involved has recently been introduced by Madlener and Stagl (2000, 2002) and Madlener (2001). Based on principles derived from ecological economics and co-evolutionary theory, they employ full fuel cycle data from life-cycle analyses as well as multiple criteria decision aid (MCDA) techniques for the design of renewable energy promotion instruments that better reflect the multiple dimensions of sustainable development and the prevailing social preferences. The latter are revealed by confronting stakeholders with the impacts of renewable energy use, measured by a number of (expert-chosen and appropriately explained) socioeconomic and biophysical indicators, whose importance they are asked to rank (e.g., by pairwise comparisons of impacts). The derived ranking of technologies can then be easily incorporated, for example, into schemes for the promotion of renewable energy utilisation that are based on economic incentives, such as guaranteed feed-in tariffs, or (the more competition-based) tradable certificate scheme as part of a (voluntary or mandatory) quota target for power from renewables.[5] In the face of market liberalisation and the current trend to increase competition, tradable 'green' certificate schemes have attracted much interest in the last years, and have been, or are planned to be, implemented in several EU member countries, several US states, and in Australia.

Concerning the relative cost of renewable energy supply options, as compared to non-renewable ones, it appears that – partly because of a lack of internalisation of external costs caused by non-renewable energy uses – renewables still tend to be relatively expensive options for energy-based GHG mitigation (Wohlgemuth and Missfeldt 2000; Rabl 2001). Moreover, calculation of the external costs of global warming caused by energy use is a complex and demanding task (as it is for some other external cost categories, such as biodiversity), and due to the long time scales involved, results are particularly sensitive to the chosen discount rate (EC 1995). Although the inherent uncertainties and irreversibilities make policy design and optimal timing challenging (Pindyck 2000), it does not, however, make renewables *a priori* an unattractive option for

GHG mitigation, provided the whole range of (net) benefits (i.e. GHG-related and ancillary benefits) accruing from their use is carefully evaluated and taken into account.

3 IMPACTS OF GLOBALISATION AND MARKET LIBERALISATION

Globalisation leads to increasing integration of national economies and increasing flows of goods, capital and know-how across regional and national borders. Closely related to globalisation is the liberalisation of markets. Within the EU, the ongoing liberalisation of the electricity and gas markets, and the implied shift from monopolistic and protected to competitive energy supply systems, has led to an enormously dynamic transition in some countries, as well as to increased risk and uncertainty as a result of the intensification in competitive pressures.[6]

The environmental impacts of the ongoing energy sector reforms in Western Europe are pulling in different directions. On the one hand, there is concern that re-regulated energy markets may not always adequately incorporate the environmental implications of resource development and energy consumption decisions. Also, it is likely that lower energy prices lead to higher demand for energy and non-energy services via the so-called 'rebound' effect (e.g., Greening et al. 2000; Binswanger 2001).[7]

On the other hand, however, the energy sector reform process also offers an opportunity to promote positive environmental changes: because the sector is already in transition, it may, in principle, become easier to tackle energy-related environmental issues. For instance, end-user choice gives electricity consumers the possibility to choose efficient energy services and green electricity.[8] In the past, policy-makers and monopolistic utilities made these choices on behalf of the consumers, whereas it is now more and more up to the energy consumer who can make responsible electricity purchase decisions (Weller 1998; Elliott 1999; Wiser 1998, 2000; Lipp 2001; Roe et al. 2001, among others).

The overall effect is not clear at all, and it critically depends on the concrete implementation of the sector reform and the particular situation of the country in question. Table 11.1 shows some likely environmental implications of different reform steps in the electricity sector.

Table 11.1: Environmental implications of power sector reforms

Reform	Environmental implication
Commerciali-sation	• Reduction of waste and better accounting (+) • Reduction of energy losses (+)
Privatisation	• Boost for off-grid renewable generation (+) • Assets upgraded by private capital (+)
Unbundling	• Distributed generation depends on ability to capture system-wide benefits (0) • Equal access to transmission depends on contract terms (0) • Energy losses reduced due to profit incentives (+)
Wholesale competition	• Low capital cost, dispatchable generation rather than renewable generation favoured by short-term-oriented markets (–) • Highly efficient fossil generation technologies favoured (+) • Non-renewable generation favoured (–)
Retail competition	• Low capital cost generation favoured (–) • Incentives to invest in environmental technology R&D reduced (–) • Retail supplier could seek competitive advantage based on environmental attributes of generation mix ('green marketing') (+)

Notes: (+) positive; (–) negative; (0) depends on concrete implementation.

Source: USAID (1998).

In a liberalised electricity market the Alps are important in at least two respects: as a natural barrier for electricity trade in the heart of Europe (which can create market distortions) and as a huge water reservoir for the renewable generation of electricity. Because product differentiation, usually by means of some sort of labelling, plays an increasingly important role in competitive power markets the latter turns Alpine regions into attractive energy suppliers on the one hand. On the other hand the economics of power generation, transmission and distribution are fundamentally changed in a restructured market environment, which – together with the significant resistance against a further expansion of large hydro-power schemes, mainly for environmental reasons, and stricter regulations on the minimum flow requirements of rivers, reservoir elevation, and ramping restrictions – tends to considerably reduce the degrees of freedom for, and increase the complexity of, optimal plant operation of existing Alpine pump storage power plants (Romerio 1998; Edwards et al. 1999). Moreover, apart from environmentally motivated reasons, current overcapacities and unsustainable pricing practices (often below marginal cost) in the European power market, as well as the high capital intensity and risks involved, tend to make investments in large hydro-power schemes in the Alps at least in the near to medium term rather unattractive.

4 GREENHOUSE GAS ABATEMENT POTENTIAL OF RENEWABLES

Global Potential

The World Energy Council (WEC) has estimated that the carbon dio-xide reductions deriving from the deployment of all renewables under an ecologically driven scenario could remove around 9 000 Mega-tonnes of carbon (MtC) from CO_2 emissions annually by 2020. This amounts to more than 40% of the current energy-related CO_2 emissions from the power sector alone (WEC 1993). In a more recent study, IIASA/WEC (1998) simulations indicate that by the year 2100 a renewables share of between 30% and more than 80% in the world's primary energy mix could be technically and economically feasible.

Table 11.2 shows typical GHG emissions per kilowatt-hour of electricity generated that result from different sources of power generation (based on life-cycle analyses).

Figure 11.1, referring to a different study, depicts the life-cycle CO_2 emissions for electricity generation from fossil fuels (coal, diesel, gas) and several renewable sources, on a per unit of generation basis. As can be seen, although exhibiting much lower GHG emissions than their fossil-fuel counterparts, several renewable energy technologies do have non-negligible CO_2 emissions (photovoltaics and at a somewhat lower level geothermal energy lead the list).

Table 11.2: GHG emissions from various sources of electricity production

Source	CO_2 equivalents (in g/kWh$_{el}$)
Hydro-power	4
Nuclear power	16
Photovoltaics	100–200
Brown coal	1 340
UCTE mix[9]	495

Source: CIPRA (1998, p.81).

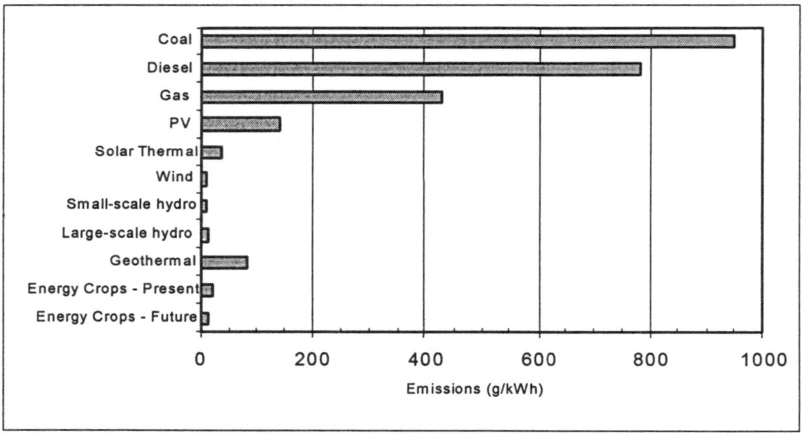

Source: © OECD/IEA (1998, p.15).

Figure 11.1: Life-cycle CO_2 emissions from fossil- and renewables-based generation of electricity

Figure 11.2 provides some insights with regard to the differences in CO_2 emissions of heating technologies, based on a full fuel chain life-cycle analysis. As can be seen, a tremendous amount of CO_2 per kWh can be saved by switching from fossil- to a bioenergy-based heating system, which today, due to impressive technology improvements, need not detract from the levels of comfort (automation) typically offered by modern oil- and gas-fired systems and nevertheless involves very low air pollutant emissions.

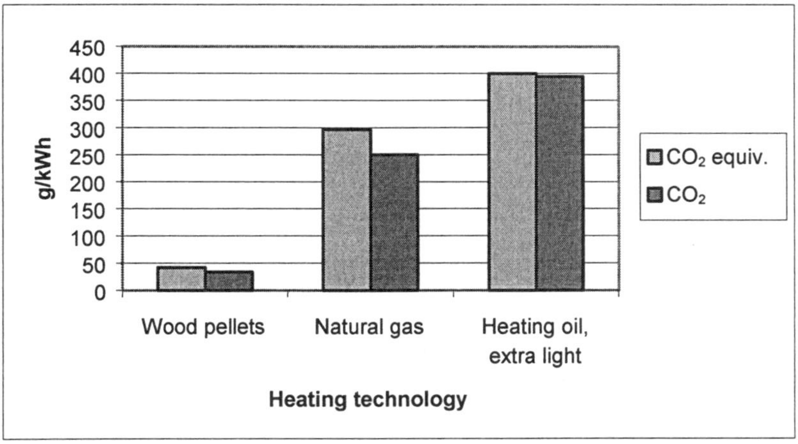

Source: Jungmeier et al. (1999).

Figure 11.2: Comparison of CO_2 emission factors of different heating technologies in Austria (residential central heating boilers)

Comparative Cost Aspects

In contrast to the adverse local and regional effects of pollutant emissions (such as urban smog and acid rain), GHG emissions have an impact on the global problem 'anthropogenic climate change'. Therefore, given spatially grossly varying marginal GHG abatement costs, a question typically raised by economists is how much GHG mitigation a particular Alpine region should achieve relative to other regions in the world (with possibly much lower or higher marginal GHG abatement), in order to come up with an optimal least-cost GHG abatement strategy. Similarly, damage arising from climate change can vary strongly not only among the different regions in the Alps, but also among nations around the world. Examples include the vulnerability to catastrophic landslides or avalanches or flooding, income dependency on certain climate-dependent industries, susceptibility to infectious diseases like Malaria, disappearance of whole islands in the Pacific Ocean because of sea level rises, and the disappearance of glaciers in the mountains. Hence some regions are exposed to much higher risks than others and, therefore, have a higher incentive to foster climate change mitigation measures.

Another line of argumentation follows the polluter-pays-principle: for the Alpine regions one could argue that, due to the fact that Alpine areas typically provide a tremendous amount of carbon neutral energy from hydro and biomass resources to other, more polluting regions, they have less of a moral obligation to reduce GHG emissions than more fossil-fuel-dependent regions. In this sense it is important in discussions to define clear system boundaries and to decide on whether the use of flexible mechanisms that allow the transfer of carbon credits and debits from one location (or entity) to another is permissible. Emission trading, which is commonly based on such comparative mitigation cost advantages and calculations (i.e. it does not account for other aspects of the GHG-emission-creating or -avoiding technologies used), could actually greatly influence the GHG mitigation actions taken in Alpine regions – including the utilisation of renewable energy sources.

5 RESOURCE, TECHNOLOGY AND SOCIAL ACCEPTANCE ASPECTS OF RENEWABLE ENERGY USE IN THE ALPS

Resources and Technologies

In the Alpine region the resources for the use of hydro, biomass, solar and wind energy are generally particularly good (for the case of photovoltaics see Bergauer-Culver and Jaeger 1998; Durisch and Bulgheroni 1999, among others).[10] Often a combination of solar collectors for hot water preparation and a biomass heating system are used for meeting the heating demands. For other energy services and in times of extreme weather conditions usually fossil-fuel supplements (especially oil and gas) are employed. Important prerequisites for sustainable energy use are the reduction of energy demand and an optimised level of energy efficiency, before remaining energy service needs are met by means of renewable energy sources (managed according to precautionary and sustainability criteria).

Hydroelectric power meets most of the electricity requirements in the Alps. Its use at the current level is widely accepted, although ecologists have claimed that the construction and operation of hydro-power schemes is often associated with significant negative local and regional impacts (Truffer et al. 2001). The construction of additional large storage and/or run-of-river power plants, while technically feasible (estimates of the potentials vary), has become less likely in the face of public opposition (see the subsection below) and the presence of a more competitive and hence much riskier market environment that tends to put capital-intensive technologies at a disadvantage. In spite of the difficulties mentioned for large-scale hydro-power plants the generating capacity could still be increased to some extent both by the construction of (environmentally sound) additional small-scale hydro-power schemes and the revitalisation and technical upgrading of existing ones which also bears a considerable potential.

Over the last decade, investment and running costs of modern *wind energy* schemes have been brought down to a level comparable with more conventional technologies, so that electricity generated by fossil and nuclear power stations could, during the winter months, be largely substituted by wind power. Whereas approximately two-thirds of hydroelectric power are generated in the summer months and only one-third during the winter, the situation is approximately reversed with regard to wind power. If favourable hillside and low-mountain areas were to be properly utilised, wind power would be ideally suited to complement run-of-river and storage power plants and to replace fossil-based alternatives. Obviously, such developments will have to take special account of the requirements of nature conservation, and the affected local population's perception of the visual and noise (and possibly other adverse) impacts.

The merits of *photovoltaic cells* appear to lie in the fact that they can be integrated into existing or newly constructed roof-tops and that the necessary components are of simple design and without movable parts (Jackson and Oliver 2000). This type of direct electricity generation from sunlight can play a significant role, even in the winter months, in Alpine locations with little winter fog. For many Alpine buildings without a grid connection it has served to reduce or even eliminate the use of diesel generators, which are often kept for back-up purposes. Grid-connected photovoltaic systems have been extensively tested and are increasingly being installed. Despite the existence of a vast number of sparsely populated areas, however, this technology is of rather limited significance in the Alpine regions, mainly because of a lack of (suitable) space and owing to the visual intrusion large-scale schemes would probably cause. Nonetheless, it has been proposed, for example, to attach photovoltaic panels on a large scale to humanmade avalanche protection units.

Some 44 percent of the Alps are covered by *forests*, bearing a massive and yet to a large extent untapped potential of biomass for *bioenergy use*. Local district heating plants can supply heat to towns and villages and – if constructed as combined heat-and-power (co-generation) plants – can help to meet the electricity requirements, especially in winter. In Austria, for example, more than 500 local district heating plants with an installed capacity of some 650 MW produce energy from woodchips (2000 figures), and all classes of heating systems based on woodchips together (approximately 25 000 systems) have an installed capacity of more than 2 000 MW today. Also, contrary to the still widespread belief that pollutant emission levels from modern wood stoves are high, these were greatly reduced over the last decade and operating comfort substantially increased (e.g., by employing semi-automatic woodchips or fully automatic wood pellet stoves).[11] Apart from forestry, sustainable agriculture can make a major contribution to the sustainable supply of energy (FAO 2000). Straw from crops is increasingly used for heating purposes. *Biogas* generated from grass, slurry and manure, clarification sludge and refuse can contribute significantly to the energy supply needs and reduce the emission of methane into the atmosphere; in addition, the fermented substrate provides excellent fertiliser for agriculture.

In Alpine regions, most of the heat for domestic hot water can be generated by *solar collectors*. In Austria, which in this respect is still at the very forefront in Western Europe, more than 1.9 million square metres of collector area have been installed so far (a figure that accounts for less than 1% of total energy supply, but is still growing quite rapidly), initiated mainly by environmentally conscious do-it-yourself groups and pushed by the existence of subsidies granted by the provincial governments (Ornetzeder 2001). The collectors also heat swimming pools, and supply heat when the weather is cool but not cold enough for the main heating system to be switched on.

Social Acceptance Aspects

The increased use of renewable energy sources is often hampered by non-technical barriers. The main challenges that can delay the realisation or lead to the failure of projects include: (i) difficulties with funding, financing and insuring; (ii) unfavourable administrative conditions; (iii) organisational difficulties; (iv) infrastructural challenges; (v) lack of knowledge; and (vi) insufficient perception and acceptance (cf. Stadlober and Hahn 1998; Roesch and Kaltschmitt 1999, among others).

Much has been done over the last few years in the financial sector, e.g., by improving the funding conditions and chances to obtain credit, or by tax incentives and tax allowances. In order to overcome administrative and organisational barriers, concerted political action across different countries, institutions and authorities and organised teamwork between the individual partners is needed. Moreover, many people still know little about energy from renewable sources and hence are often prejudiced to some degree. Several renewable energy technologies also suffer from image problems – particularly those based on biomass – and are still frequently considered as low-tech, old-fashioned, inefficient, unreliable, and more difficult and time consuming to operate than traditional ones based on oil or gas or on electricity (Rakos 1999, p.59). Besides, many people are still not aware of the CO_2 neutrality of biomass for energy use from sustainably managed forests, in contrast to the use of fossil fuels. Similarly, the population in the Alpine region has increasingly signalled resistance against additional dams and large-scale pump-storage power stations.

For the above-mentioned reasons it is necessary to correct any existing biases and to achieve a better understanding and acceptance by providing stakeholders with adequate and objective information on new technologies and their respective (relative) benefits. The focus should therefore be placed on information, for instance, about the net reduction in GHG emissions, the possibilities of maintaining existing and of creating new jobs and income, or the higher economic value and other benefits derived from productive use of set-aside land for growing crops or grass for bioenergy. The timing of information and its targeting at an adequate level, as well as the facilitation of an open and independently and professionally coached discussion involving the relevant stakeholders (including NGOs and regional policy-makers), are important ingredients in the successful promotion of renewable energies.

6 THE CASE OF THE ALPINE PROVINCE OF CARINTHIA

General Overview

Carinthia is the most southern of the nine Austrian provinces, and lies fully within the Alpine region (Figure 11.3). It covers an area of 9 533 square kilometres and is the home of about 570 000 people. It shares its borders with Slovenia (170 km), Italy (109 km) and three other Austrian provinces (Styria, Tyrol, Salzburg). Agriculture and forestry account for some 2% of gross value added, manufacturing industry for 19%, trade and transport 18%, tourism 6.5%, private services 18.5% and public and other services about 22% (1997 figures).

In 1997 primary energy production in Carinthia amounted to 31 466 TJ (AKL 1999). The indigenous supply structure is dominated by hydropower (50.3%), forest residues (26.4%), other biofuels (21.1%) and others (2.2%). In order to match total energy needs, 45.5% of the overall energy supply have been imported from other Austrian provinces, 18% from abroad (1997 figures). Total energy exports accounted for 13 877

TJ, of which 71.4% was exported to other Austrian provinces and 28.6% abroad.

On the demand side, gross domestic energy consumption in Carinthia was some 68 943 TJ, split into the consumption of petrol fuels (25.6%), non-hydro renewables (23.5%), hydro-power (19%), heating oil (18.8%), gas (7.9%) and coal (5.3%). The category non-hydro renewables is composed of forest residues (54.8%), other biofuels (41%), heat pumps and solar thermal (3.2%) and combustible wastes (0.9%). In other words, the contribution of renewable energy sources to overall primary energy consumption is around 42.5% (AKL 1999; see Faninger 1993 for the historical development). Split by energy end-use sectors the private households accounted for 39.6%, industry for 28%, the service sector for 13.6%, agriculture for 10.4% and transportation for 8.5% (1996 figures).

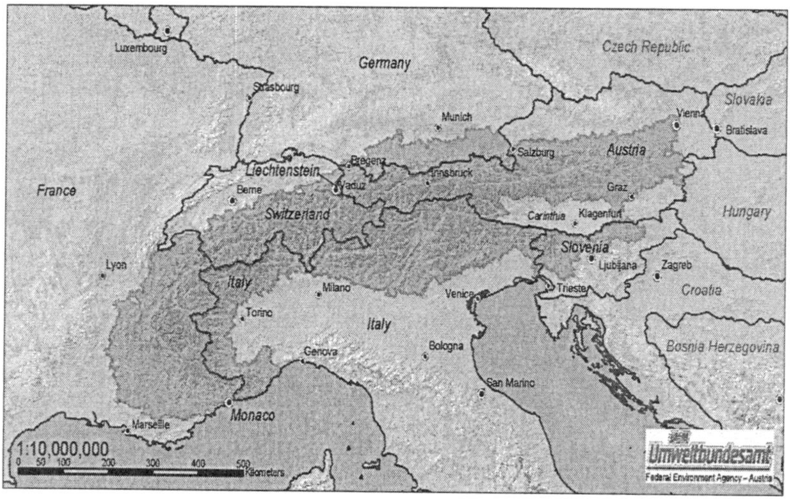

Source: Courtesy of K. Placer, Federal Environment Agency Austria.

Figure 11.3: Location of the Austrian province of Carinthia (Kaernten) within the Alpine region, as defined in the Alpine Convention

Some Results from a Scenario-based Modelling Approach

The results of a study on the future market penetration of renewables in Carinthia, using the SAFIRE model[12] and a scenario-based approach, indicate that an increase in the use of renewable energy not only has potential to mitigate GHG and pollutant emissions, but can also contribute to the economic and social dimension of sustainable development in the province (Verbundplan/ESD 1997; Wohlgemuth 1998). In particular, simulation runs have shown that the cost-benefit indicators calculated by the model (e.g., employment, government revenues, CO_2 and pollutant emissions) indicate conflicting results: with respect to 'market penetration', 'employment' and 'investment', a scenario based on *subsidising renewable energy* provides the preferable overall option. In contrast, another scenario based on the imposition of a harsher *fossil-fuel tax* regime revealed that the indicators for 'government revenues', 'CO_2 emissions' and 'import dependency' move in a desirable direction. In the practical context of energy and environmental policy design this implies that strong policy trade-offs can be expected.

In the study mentioned, the *technical potential*[13] of renewable energy utilisation in Carinthia has been estimated to be around 13 000 GWh (current total final energy consumption is around 18 000 GWh).[14] Roughly 50% of this potential can be attributed to large-scale hydro-power and slightly less than 30% to biomass in the form of forest

Table 11.3: Estimated technical potential of renewables for energy in Carinthia, 1994–2020

Technology Year	1994	2000	2010	2020
		(in GWh p.a.)		
Wind	65	65	65	65
Small hydro	560	560	560	560
Large hydro	6 440	6 440	6 440	6 440
Photovoltaics	241	251	267	278
Active solar thermal	409	426	454	473
Passive solar thermal	469	482	509	532
Cooling/daylighting	19	20	21	22
Forest residues	3 679	3 679	3 679	3 679
Energy crops for ethanol	371	402	439	458
Energy crops for biodiesel	223	241	263	274
Municipal waste	378	378	378	378
Landfill gas	5	5	5	5
Total	12 858	12 950	13 081	13 165

Note: Small-scale hydro-power plants in SAFIRE are those with a declared net capacity of less than 5 MW.

Source: Verbundplan/ESD (1997).

residues (Table 11.3). Only about half of the technical potential is currently utilised. Due to reasons already outlined in section 3, public resistance against a further expansion of hydro-power schemes and the various risks and competitive pressures imposed on the plant owners by market liberalisation make bioenergy options a more likely candidate for providing a significant future contribution (and hence share) to the province's fuel mix.

The *market penetration* of individual renewable energy technologies under the three scenarios analysed ('Reference', 'Subsidies' and 'High tax') can be seen from Table 11.4. The results indicate that although the total market penetration is predicted to only increase modestly (Reference: 0.2% p.a.; Subsidies: 0.5% p.a.; High tax: 0.3%), mainly due to the assumed constant technical potential for the most prominent

Table 11.4: Renewable energy market penetration in Carinthia, 2000–2020, different scenarios

Scenario	Reference		Subsidies		High tax	
Year	2000	2020	2000	2020	2000	2020
Technology	(in GWh p.a.)					
Wind	3.5	3.5	3.5	3.5	3.5	3.5
Small hydro	364.9	364.9	364.9	364.9	364.9	364.9
Large hydro	4 198.0	4 198.0	4 198.0	4 198.0	4 198.0	4 198.0
Photovoltaics	0.1	0.1	0.1	0.1	0.1	0.1
Active Solar thermal	38.5	40.3	47.6	121.1	40.7	74.6
Passive Solar thermal	8.4	155.9	13.1	162.2	8.3	149.8
Cooling/ daylighting	3.3	20.0	5.2	20.3	3.3	19.2
Forest residues	1 732.7	1 779.6	1 743.7	1 821.7	1 732.7	1 793.7
Energy crops for ethanol	0.0	0.0	0.0	197.4	0.0	0.0
Energy crops for biodiesel	0.0	0.0	0.0	92.2	0.0	71.1
Municipal waste	26.8	57.7	32.7	48.2	25.5	46.8
Landfill gas	4.4	4.8	4.8	4.8	4.8	4.8
Total	6 380.6	6 624.8	6 413.5	7 035.4	6 381.8	6 726.6

Source: Verbundplan/ESD (1997).

resources, hydro-power and forest residues, certain technologies (e.g., passive solar thermal, photovoltaics) are expected to exhibit substantial increases, though from a rather low base year level.

In all three scenarios considered, total *CO_2 emissions* increase as a result of the assumed large growth in energy demand. In the base year (1994), SAFIRE calculates the total annual CO_2 emissions in the energy sector at 3.19 million tonnes (Mt). For the purpose of comparison, Table 11.5 depicts the development of the energy-related CO_2 emissions in Carinthia during the 1990s according to the official statistics.

Figure 11.4 shows the emissions in the 'Reference' and 'Subsidies' cases. By 2020, without induced additional market penetrations (for simplicity referred to as 'no penetration' in the graphs), total emissions are predicted to increase by 22% to 3.9 Mt of CO_2. The additional penetration of renewables in the two alternative cases reduces the effect, but not to such an extent as to lead to an absolute decrease in emissions. Owing to the larger renewables penetration, the 'best' outcome in terms of CO_2 mitigation is in the 'Subsidies' strategy, where 0.3 Mt of CO_2 are saved, leaving a total of 3.6 Mt in 2020. On the renewables side, the largest effect is caused by growing crops for ethanol due to the carbon-neutral emissions of this fuel. On a similar basis, biodiesel production and forest residues consumption come third at approximately 22 000 tonnes saved from each. The other main improvements are caused by the implementation of energy-saving technologies.

The 'High tax' strategy shows a different story, as is shown in Figure 11.5. Owing to the lower energy demand growth, total CO_2 emissions in Carinthia reach a total of just below 3.68 Mt by 2020 in a 'no (additional) penetration' situation. This figure is only marginally greater than the total figure under the 'Subsidies' strategy. The tax-induced renewable energy penetration in the 'High tax' strategy reduces the 'no (additional) penetration' figure for CO_2 emissions by approximately 0.25 Mt to 3.44 Mt in 2020.

Table 11.5: Energy-related CO_2 emissions in Carinthia, 1993–1997

Year	1993	1994	1995	1996	1997
CO_2 (mio. t p.a.)	3.07	3.06	3.09	3.14	3.13

Source: AKL (1999, p.21).

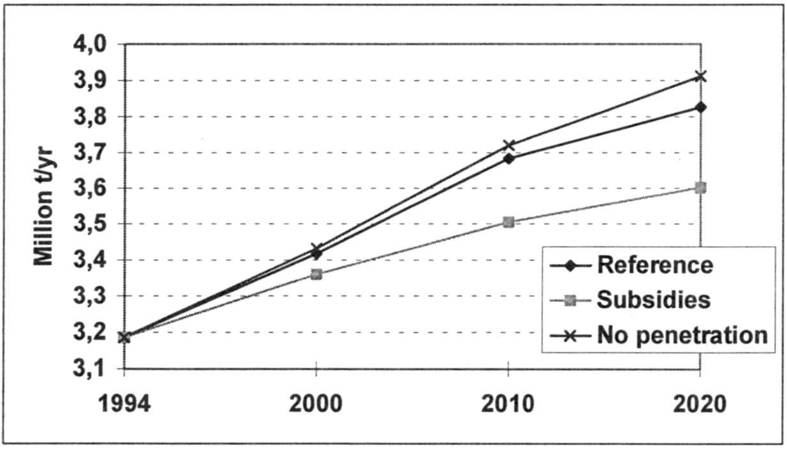

Source: Verbundplan/ESD (1997).

Figure 11.4: Total CO_2 emission scenarios for Carinthia, 1994–2020 (high energy growth)

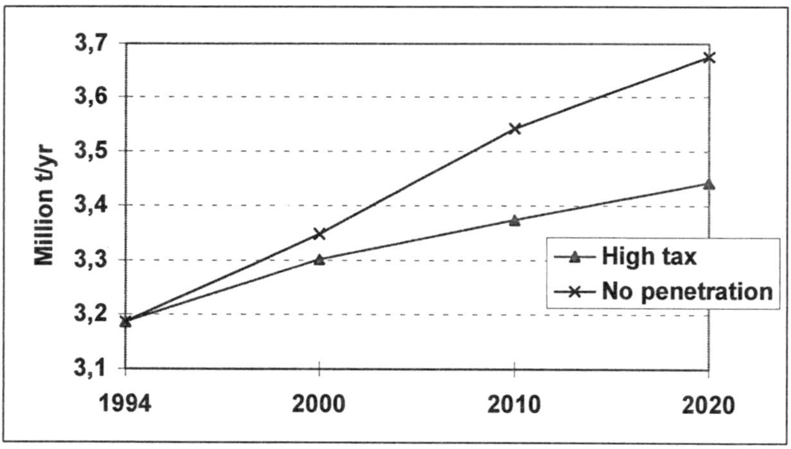

Source: Verbundplan/ESD (1997).

Figure 11.5: Total CO_2 emission scenarios for Carinthia, 1994–2020 (low energy growth)

The main conclusion that can be drawn from this study is that for Carinthia total CO_2 emissions can best be mitigated by implementing energy-saving strategies that minimise the increase in total energy demand, as is shown in the 'High tax' strategy. There is a larger effect on emissions for the 'Subsidies' strategy, but this does not bring it below the totals for the low-energy growth scenario. However, the assumed tax increases are not sufficient to cause an absolute decrease over the time period of the study. Therefore, provided higher tax increases are not considered feasible, other methods, such as the introduction of energy efficiency measures (e.g., technology or performance standards; research, development and demonstration – RD&D) or energy saving measures (e.g., demand-side management, integrated resource planning) will be needed to stabilise and eventually reduce CO_2 emissions to a sustainable level.

Renewable Energy Promotion Policy in Carinthia

The pillars of the current energy policy in Carinthia rest on an increase in energy efficiency and the promotion of renewable energy use by grants, information and education – rather than on command-and-control measures (AKL 2000). During the past few years, the Carinthian government has subsidised the use of renewable energy technologies, particularly heat pumps, solar thermal collectors, heating systems based on firewood and forest residues (single stove) and the erection and extension of district heating systems.

Regulations in force since January 1996 subsidise connections to district heating systems based on renewable energy sources, waste heat or co-generation, provided at least 80% of the annual energy generation is from one of these sources. Public and industrial investors can get a subsidy of up to 20% of the actual investment.

The Carinthian government also subsidises the installation of solar thermal collectors and biomass heating systems. The maximum subsidy granted may not exceed 20% of the total investment costs. The Austrian energy agency E.V.A. provides a useful overview of the subsidies provided by the nine Austrian provinces for electricity produced from renewable energy sources and fed into the grid (www.eva.ac.at/enz/einspeis_at.htm), differentiated by technology and by time of use (summer/winter and day/night, respectively). According to the revised tariffs in preparation for Carinthia at time of writing this chapter, electricity generation from photovoltaics, for example, would be subsidised via a guaranteed feed-in tariff of Euro (EUR) 0.7270 per kWh, while solid biomass-based electricity is paid between EUR 0.0727 to

EUR 0.1745 (depending on plant size and type of biomass, with forest residues receiving the highest tariff), power from liquid biomass between EUR 0.0727 and EUR 0.1450 (depending on the type of liquid; plant oil and biodiesel receive the highest tariffs), power from gaseous biomass between EUR 0.0900 and EUR 0.1600 (electricity from agricultural biogas would receive the highest tariffs), geothermal power EUR 0.0875 and wind power EUR 0.1075 per kWh.

The quota-based tradable certificate scheme for the promotion of small-scale hydro-power plants (≤ 10 MW; Madlener and Drillisch 2002) that was introduced with the amended Austrian electricity act (ElWOG 2000; in force since 1 October 2001) has also been expected to have some significant implications for Carinthia, which ranks fifth among the nine Austrian provinces in the power generation from small hydro-power stations (1999 share of total production in Austria: 10.8%). Unfortunately, the scheme was ill-designed and abandoned in July 2002, in the course of a renewable energy promotion harmonisation effort among the nine provinces, before it was ever fully operational.

Finally, it should be mentioned that apart from direct (e.g., subsidies, tax credits) and indirect (e.g., RD&D spending) promotion schemes for renewable energy production ('second-best' options in economics terminology), a tax on CO_2 emissions (or alternatively a carbon-based energy tax) can also play a key role in ensuring a higher degree of market penetration for renewable energy sources. If set at an optimal level, such a tax would constitute a 'first-best' solution (internalisation of the climate-change related external costs and thus rectification of market failure), but is politically often not feasible – which is why second-best instruments are in fact so popular.

7 CONCLUSIONS

In this chapter we have studied the connections between climate-change mitigation and the use of renewable energy sources for heat and electricity in the Alps, exemplified by a case study on the Austrian province of Carinthia. We find that sustainable energy use is just one of many goals towards a more sustainable development in the Alpine regions, and that the use of renewable energy sources is just one possible and important (albeit not always the most cost-efficient) approach to tackling the severe and long-term problem of global warming.

Anthropogenic climate change is clearly a global problem with a local impact and local causes. In the case of the Alpine regions, due to the greater ecological, social and economic sensitivity of the area, and the important cultural heritage of the Alps, decision-makers can in some sense be expected to be more willing to take earlier and more decisive action than in other regions. The protocols developed under the framework of the *'Alpenkonvention'* are indeed encouraging in this respect, although some probably still have a long way to go before the ratification process by the national parliaments is completed. Policies taken in the Alpine regions may also serve as an important model for other socioeconomic-ecologically fragile areas both in the developed and in the developing world.

In our opinion the fact that Alpine systems are more fragile and vulnerable makes the inhabitants more receptive in adopting measures against climate change. Often climate change is one of several other adverse factors threatening the sustainable development of Alpine regions. In the energy sector the need to improve sustainability is felt particularly strongly, due to the structural changes induced by the ongoing liberalisation efforts towards a single European market, where cost efficiency, harmonisation of the rules of the game and competition are gaining importance. Innovative technologies, like fuel cells and microturbines, especially when powered from renewable sources, offer some new and interesting options towards a more sustainable development, but unless unsustainable absolute levels of consumption are curbed, many of the most pressing problems are not going to be resolved.

The case study of the Austrian province of Carinthia has shown that under current conditions and without the imposition of severe additional policies, the diffusion of renewable energy sources alone is unlikely to lead to a stabilisation, or even reduction, of the prevailing GHG emissions levels in the Alpine area. Determined policies and smart policy instruments that are able to steer consumption patterns in a more sustainable direction, plus large-scale substitution of less by more energy-efficient technologies, and fossil and nuclear by renewable fuels, as well as the avoidance of excessive rebound effects, are important keys to long-term success.

The challenge ahead, therefore, will be to find the fine and twisted line between an appropriate consideration of the many particularities of very sensitive areas like the Alpine regions on the one hand, and European-wide consolidation efforts aiming at higher economic and energy efficiency, a level playing field for competition, welfare and distributional equity on the other hand. The growing awareness and understanding of the manifold issues involved will help to identify pathways, and intelligent measurement and promotion systems, for a more sustainable development.

NOTES

1. See, for example, the Kyoto Protocol (UNFCCC 1997, especially Article 2). The Intergovernmental Panel on Climate Change (IPCC) has acknowledged the paramount role renewable energy is expected to play in meeting the ultimate goal of virtually replacing fossil fuels, noting that 'in the longer term, renewable sources of energy could meet a major part of the world's demand for energy' (IPCC 1996, p.589).
2. See also EC 2001a for the latest proposal of an EU Directive on the internal energy market.
3. Chapter 13 includes two programme areas to further elaborate the problem of fragile mountain ecosystems. These are: (1) generating and strengthening knowledge about the ecology and sustainable development of mountain ecosystems; and (2) promoting integrated watershed development and alternative livelihood opportunities.
4. The eight other protocols, all of which are relevant for a sustainable development of the Alpine region, deal with: (i) Nature Protection and Care; (ii) Mountain Agriculture; (iii) Regional Planning and Sustainable Development; (iv) Mountain Forests; (v) Tourism; (vi) Transportation; (vii) Soil Protection; and (viii) Dispute Settlement (within the Alpine Convention framework).
5. These certificates are sometimes also referred to as 'tradable green certificates' and the underlying mechanism as 'Renewable Portfolio Standard (RPS)'; see for example Espey (2001), Morthorst (2000) and Rader (2000), among many others.
6. The European Union's Directive Governing the Common Rules for the Internal Electricity Market (EC 1997a) is based on the same understanding of competition as is applied in the Directives for the telecommunication market, and other grid-based fuels, such as natural gas. The theoretical basis for this understanding is the model of contestable markets. In order to apply this model, two fundamental preconditions have to be met: there must not be a natural monopoly, and market entry and exit must be free. As a consequence of the understanding of competition based on contestability, the EU in its Electricity Directive requires a vertical disintegration of the electricity supply industry (ESI) into electricity production, transmission and distribution and final consumer services (unbundling). The intention is to remove market power (and its associated rents) and cross-subsidies.
7. In addition, a market-based electricity price setting can also be expected to result in increased price volatility (Amundsen and Tjøtta 1997).
8. Green pricing is a concept in which consumers voluntarily agree to pay a higher price for electricity generated by environmentally friendly and, particularly, renewable energy sources. Surveys carried out in many developed countries repeatedly indicate a willingness by electricity consumers to pay a higher price for clean energy, and green pricing provides customers with a choice to do precisely that. Being a 'market-driven' approach to environmental protection, green pricing is likely to receive increased emphasis in liberalised markets; and in fact, as one of the few non-price means of distinguishing one's service in a commodity market, green energy can be expected to become an important marketing strategy in competitive market environments (Wuestenhagen 1998).

9. The term 'UCTE mix' refers to the relative GHG emissions caused by the power generation mix from which electricity is transmitted through the networks of the members of the 'Union for the Co-ordination of Transmission of Electricity' (UCTE). UCTE co-ordinates the interests of the transmission system operators in 20 European countries. Their common objective is to guarantee the security of operation of the interconnected power system. Through the networks of the UCTE, 400 million people are currently supplied with electric energy. The annual electricity consumption totals approximately 2.100 TWh.
10. Based on a computer simulation for two spots in the Austrian Alps (the town of Leonding and the summit Loser), Bergauer-Culver and Jaeger (1998) have studied the output that can be expected from photovoltaic power plants in Alpine regions. PV systems installed at higher altitudes usually offer higher outputs, because of clearer sky, snow reflection, lower ambient temperatures and better cooling of the panels by the wind. The study by Durisch and Bulgheroni (1999), focusing on the Swiss Alps situation, additionally covered the feasibility of solar thermal systems, sun-tracking PV panels and the influence of humidity and rainfall measurements and impacts.
11. This applies particularly to pellet heating systems, which can be operated fully automatically and which, due to the homogeneity of the fuel, exhibit very low pollutant emission levels.
12. For a description of SAFIRE (Strategic Assessment Framework for the Implementation of Rational Energy) see Madlener (2000) or Madlener and Myles (2000, Section 3.10), and the references therein. SAFIRE has been part-funded by the European Commission under the JOULE programme, and recently has been used for the determination of the indicative targets contained in the EU Renewables Directive. It is an engineering-economic bottom-up model for the assessment of first-order impacts of 'rational' (i.e. renewable and new non-renewable) energy technologies on a national, regional or local level against a background of different policy instruments and scenario assumptions. It is a framework that consists of a database and a computer model that provides decision-makers with a tool to evaluate the market and impact of new energy technologies and policies. Recently, SAFIRE has been updated to take into account the calculation of baselines within the Kyoto framework. The time horizon of the current model version is 36 – 40 years, depending upon the base (starting) year specified by the user. Outputs are available for a number of marker years. SAFIRE is data-driven; the software includes an extensive database for 22 renewable energy technologies, eight new non-renewable energy technologies and seven fuelling options for co-generation plants (which is assumed to be all decentralised) including fuel cells. Also, it has a domestic sector demand-side management option. The range of countries covered within SAFIRE include the EU-15, four other Western European countries (Cyprus, Iceland, Norway, Switzerland), 12 Central and Eastern European countries and eight other countries worldwide (Brazil, Canada, China, India, Indonesia, Japan, Mexico, USA). Further information may be obtained from Mark Whiteley (mark@esd.co.uk), ESD Ltd., Overmoor, Neston, Corsham SN13 9TZ, United Kingdom.

13. The assessment of the technical potential (TP) in the cited study is based on an estimate for the energy a technology could supply if there were no economic constraints, market barriers or competition from other sources of energy. It may either be demand-constrained (e.g., as in the case of solar power) or supply-constrained (e.g., as in the case of forest residues). For some technologies covered, the TP is allowed to vary over time, in order to reflect (a) policy changes (wastes), (b) technological progress (wind) or (c) demand levels for space heating, space cooling and day-lighting (passive solar design).
14. Note that in the study mentioned municipal waste is considered to be a 'renewable' source of energy (see Tables 11.3 and 11.4).

BIBLIOGRAPHY

Abbasi, S.A. and N. Abbasi (2000), 'The likely adverse environmental impacts of renewable energy sources', *Applied Energy*, **65** (2), 121–144.

AKL (1999), *Energiebericht 1999 fuer das Bundesland Kaernten*, Klagenfurt: Amt der Kaerntner Landesregierung (AKL), Abteilung fuer Wasser-, Abfall- und Energierecht, December.

AKL (2000), *Foerderungsrichtlinien des Energiereferates*, Klagenfurt: Amt der Kaerntner Landesregierung (AKL), Abteilung fuer Wasser-, Abfall- und Energierecht, January.

Amundsen, E.S. and S. Tjøtta (1997), 'Trade and price variation in an integrated European power market', *Applied Economics*, **29** (6), 745–757.

Bergauer-Culver, B. and C. Jaeger (1998), 'Estimation of the energy output of a photovoltaic power plant in the Austrian Alps', *Solar Energy*, **62** (5), 319–324.

Binswanger, M. (2001), 'Technological progress and sustainable development: what about the rebound effect?', *Ecological Economics*, **36** (1), 119–132.

CIPRA (1995), *Energiekonzepte im Gebirge*, Gschwandtl, I. (ed), Proceedings of the CIPRA Jahresfachtagung in Innsbruck 10 June 1994, Vienna: CIPRA Austria.

CIPRA (1998), *Energiezukunft Alpen. Die Oeffnung der Strommaerkte und die Folgen fuer die Berggebiete* (Energy Future Alps. The Opening of the Power Markets and the Consequences for Mountain Regions; in German), Proceedings of the CIPRA Annual Meeting in Locarno/ Switzerland, 22–24 October 1998, Schaan/Principality of Liechtenstein: CIPRA – Internationale Alpenschutzkommission.

Dincer, I. and M.A. Rosen (1999), 'Energy, environment and sustainable development', *Applied Energy*, **64** (4), 427–440.

Durisch, W. and W. Bulgheroni (1999), 'Climatological investigation for solar-power stations in the Swiss Alps', *Applied Energy*, **64** (4), 411–415.

EC (1995), *ExternE. Externalities of Energy*, Vols. 1–6, Luxembourg: European Commission, DG-XII (Science, Research and Development).

EC (1997a), 'Directive 96/92/EC of the European Parliament and of the Council governing the common rules for the internal electricity market', OJ L027/20, 30 January 1997, Brussels: European Commission.

EC (1997b), 'Energy for the Future: Renewable Sources of Energy. White Paper for a Community Strategy and Action Plan', COM(97) 599 final, 26 November 1997, Brussels: European Commission.

EC (1998), 'European Parliament and Council Directive 98/30/EC of 22 June 1998 concerning common rules for the internal market in natural gas', OJ L 204, 21 July 1998, Brussels: European Commission.

EC (2000), 'Bringing our needs and responsibilities together – integrating environmental issues with economic policy', COM(2000) 576 final, 20 September 2000, Brussels: European Commission.

EC (2001a), 'Proposal for a Directive of the European Parliament and of the Council of 13 March 2001 amending Directives 96/92/EC and 98/30/EC concerning common rules for the internal market in electricity and natural gas', COM(2001) 125 final, 13 March 2001, Brussels: European Commission.

EC (2001b), 'Directive 2001/77/EC of the European Parliament and of the Council of 27 September 2001 on the promotion of electricity produced from renewable energy sources in the internal electricity market', OJ L 283/33, 27 October 2001, Brussels: European Commission.

Edwards, B.K., Flaim, S.J., and R.E. Howitt (1999), 'Optimal provision of hydroelectric power under environmental and regulatory constraints', *Land Economics*, **75** (2), 267–283.

Elliott, D. (1999), 'Prospects for renewable energy and green energy markets in the UK', *Renewable Energy*, **16** (4), 1268–1271.

ElWOG (2000), Elektrizitaetswirtschafts- und -organisationsgesetz (Austrian electricity act), BGBl. I Nr. 121/2000 (Energieliberalisierungsgesetz), Artikel 7, Vienna.

Espey, S. (2001), 'Renewables portfolio standard: a means for trade with electricity from renewable energy sources', *Energy Policy*, **29** (7), 557–566.

Faninger, G. (1993), *Erneuerbare Energietraeger fuer Kaernten: Einsatzbereiche, Verbrauchsentwicklung, Umweltrelevanz und Potentialabschaetzung* (Renewables for Carinthia: Uses, Consumption Development, Environmental Relevance, and Potential Estimate; in German), Seibersdorf: Austrian Research Centre Seibersdorf (ARCS).

FAO (2000), *The energy and agriculture nexus,* Environment and Natural Resources Working Paper No. 4, Rome: Food and Agriculture Organization of the United Nations.

Flavin, C. and S. Dunn (1998), *Climate of Opportunity: Renewable Energy after Kyoto*, Renewable Energy Policy Project Issue Brief No. 11, July, Washington DC: REPP (www.repp.org).

Greening, L.A., D.L. Greene and C. Difiglio (2000), 'Energy efficiency and consumption – the rebound effect – a survey', *Energy Policy*, **28** (6–7), 389–401.

Hein, W. (1996), 'Climate and Energy', in *Green Paper on the Alps. The Alps – Touchstone for Europe*, Vienna/Berne: Friends of Nature International.

IIASA/WEC (1998), *Global Energy Perspectives*, N. Nakicenovic, A. Gruebler and A. McDonald (eds), Cambridge/New York: Cambridge University Press.

IPCC (1996), *Climate Change 1995 – Impacts, Adaptations, and Mitigation of Climate Change: Scientific-technical Analyses,* Contribution of Working Group II to the Second Assessment Report of the Intergovernmental Panel on Climate Change, R.T. Watson, M.C. Zinyowera, and R.H. Moss (eds), Cambridge/New York: Cambridge University Press.

Jackson, T. and M. Oliver (2000), 'The viability of solar photovoltaics', *Energy Policy*, **28** (14), 983–988.

Jungmeier, G., L. Canella and J. Spitzer (1999), 'Greenhouse Gas Emissions of Bioenergy Systems Compared to Fossil Energy Systems. Methodology and First Results', *Proceedings of the 4th Biomass Conference of the Americas*, Oakland, Col., 29 Aug–2 Sept 1999, Vol. I, 437–444.

Lipp, J. (2001), 'Policy considerations for a sprouting UK green electricity market', *Renewable Energy*, **24** (1), 31–44.

Madlener, R. (2000), 'Modelling the prospects for renewable and new non-renewable energy technologies in the UK and some of the consequences implied', G. MacKerron and P. Pearson (eds), *The International Energy Experience. Markets, Regulation and the Environment*, London: Imperial College Press, Ch. 18, 245–259.

Madlener, R. (2001), 'How to maintain competition and diversity? A socio-ecological-economic assessment of bioenergy options with a focus on CHP', J. Domac and K. Richards (eds), *Proceedings of the Workshop 'Socio-economic aspects of bioenergy systems: Challenges and opportunities'*, 28-31 May 2001, Rocky Mountain Region, Alberta, Canada; Zagreb: IEA Bioenergy Task 29/Energy Institute 'Hrvoje Pozar', October, 71–75.

Madlener, R. and J. Drillisch (2002), 'Tradable Certificate Schemes for Single Renewable Electricity Technologies: The Case of Small-Scale Hydro Power Promotion in Austria', *Proceedings of the 25th Annual Conference of the International Association for Energy Economics*, Aberdeen, Scotland, 26–29 June 2002.

Madlener, R. and H. Myles (2000), 'Modelling Socio-Economic Aspects of Bioenergy Systems: A survey prepared for IEA Bioenergy Task 29', September (www.iew-bioenergy-task29.hr/Task29/pdf/Models.pdf).

Madlener, R. and S. Stagl (2000), 'Guaranteed feed-in tariffs vs. tradable renewable electricity certificates: an ecological-economic comparison', *Proceedings of the 3rd Biennial Conference of the European Society of Ecological Economics (ESEE)*, Vienna, 3–6 May 2000, Paper No. 295 (www.wu-wien.ac.at/esee2000).

Madlener, R. and S. Stagl (2002), 'Sustainability-oriented promotion of renewable electricity generation', *Ecological Economics*, (forthcoming).

Morthorst, P. E. (2000), 'The development of a green certificate market', *Energy Policy*, **28** (15), 1085–1094.

OECD/IEA (1998), *Benign Energy? The Environmental Implications of Renewables*, Paris: International Energy Agency/OECD.

Ornetzeder, M. (2001), 'Old technology and social innovations. inside the Austrian Success story on solar water heaters', *Technology Analysis and Strategic Management*, **13** (1), 105–115.

Pindyck, R. S. (2000), 'Irreversibilities and the timing of environmental policy', *Resource and Energy Economics*, **22** (3), 233–259.

Price, M., T. Kohler, T. Wachs, and A. Zimmermann (eds) (2001), *Mountains of the World – Mountains, Energy and Transport*, Prepared by Mountain Agenda for the Commission on Sustainable Development (CSD), Berne: Centre for Development and Environment (CDE), Institute of Geography, University of Berne, Switzerland.

Rabl, A. (2001), 'The importance of external costs for the competitiveness of renewable energies', *International Journal of Global Energy Issues*, **15** (1 – 2), 49–58.

Rader, N. (2000), 'The hazards of implementing renewables portfolio standards', *Energy and Environment*, **11** (4), 391–405.

Rakos, C. (1999), 'Domestic heating – the biomass challenge', *Renewable Energy World*, **2** (5), September, 54–61.

Roe, B., M.F. Teisl, A. Levy and M. Russell (2001), 'US consumers' willingness to pay for green electricity' *Energy Policy*, **29** (11), 917–925.

Roesch, C. and M. Kaltschmitt (1999), 'Energy from biomass – do non-technical barriers prevent an increased use?', *Biomass and Bioenergy*, **16** (5), 347–356.

Romerio, F. (1998), 'Models of Hydropower Resource Management in the Swiss Alps', *21st Conference of the International Association for Energy Economics (IAEE)*, Quebec, Canada, 115–124.

Stadlober, M. and B. Hahn (1998), *Soziale Akzeptanz von Windkraftanlagen in Oesterreich* (Social Acceptance of Wind Power Plants in Austria; in German), study on behalf of the Austrian Federal Ministry of Science and Transportation, St. Poelten/Austria, August.

Stanford, A. (1997), 'A vision for a sustainable energy future', *Renewable Energy*, **10** (2–3), 417–422.

Truffer, B., J. Markard, C. Bratrich and B. Wehrli (2001), 'Green electricity from Alpine hydropower plants', *Mountain Research and Development*, **21** (1), 19–24.

UNCED (1993), *Report of the United Nations Conference on Environment and Development, Rio de Janeiro, 3–14 June 1992*, Vol. I, Resolutions Adopted by the Conference, Agenda 21, Document A/Conf. 151/26/Rev.1., New York: United Nations.

UNFCCC (1997), *Kyoto Protocol to the United Nations Framework Convention on Climate Change*, FCCC/CP/1997/7/Add.1, Bonn: United Nations Framework Convention on Climate Change.

USAID (1998), *Promoting Energy Efficiency in Reforming Electricity Markets: A Guidebook for Stakeholders,* prepared by Hagler Bailly Services Inc., San Francisco/CA, Report No. 98-04, March.

Verbundplan/ESD (1997), *SARI – Carinthia. Systems Approach to Renewable Integration*, study on behalf of the European Commission DG-XVII (Energy) within the ALTENER program, Final Report, March.

WEC (1993), *Renewable Energy Resources: Opportunities and Constraints 1990–2020*, London: World Energy Council.

WEC (1998), *The Benefits and Deficiencies of Energy Sector Liberalisation*, London: World Energy Council.

Weller, T. (1998), 'Green Pricing: kundenorientierte Angebote der Elektrizitaetswirtschaft' (Green Pricing: customer-oriented offers of the electricity industry; in German), *Zeitschrift fuer Energiewirtschaft*, **22** (1), 58–70.

Wiser, R. H. (1998), 'Green power marketing: increasing customer demand for renewable energy', *Utilities Policy*, **7** (2), 107–119.

Wiser, R. H. (2000), 'The role of public policy in emerging green power markets: an analysis of marketer preferences', *Renewable and Sustainable Energy Reviews*, **4** (2), 177–212.

Wohlgemuth, N. and F. Missfeldt (2000), 'The Kyoto Mechanisms and the prospects for renewable energy technologies', *Solar Energy*, **69** (4), 305–314.

Wohlgemuth, N. (1998), 'The Role of Renewable Energy Technologies in a Systems Approach: An Effective Promotion Strategy under Multiple Objectives Considerations', *Proceedings of the International Conference 'Biomass for Energy and Industry'*, Wuerzburg, Germany, 8–11 June 1998, 1117–1120.

Wuestenhagen, R. (1998) 'Green electricity in Switzerland: a mere eco-niche or the first step towards a sustainable energy market at large?', 4th European Conference of the International Association for Energy Economics/Gesellschaft fuer Energiewissenschaft und Energiepolitik (IAEE/GEE) *'Energy Markets: What's New?'*, Berlin, 9–10 September 1998, 143–151.

12. Reducing the Global and Local Environmental Impact of Transport in the Alpine Region[1]

Bertrand Chateau, Birgit Friedl, Alberto Frondaroli, Mario Keller, Romain Molitor and Karl W. Steininger

1 THE ALPINE REGION, A VERY SENSITIVE AREA

The Alps are an outstanding example of a region where ecosystems and the environment are very vulnerable. This vulnerability is partly due to the topographical and climatic situation in the Alps: The shape of the valleys (V- or U-shape) and meteorological particularities (tendency towards inversions) impede the escape of emissions from combustion processes. Consequently, air pollution along these valleys is often as bad as in urban areas and ozone smog reaches high levels. Alpine forests, which protect the settlements against avalanches and landslides, are being damaged. Noise nuisances are extremely intensified by the topography of the valleys. Pollution and noise not only have negative impacts on the environment, but also cause high health risks. On looking at sources, we find that transport is the major contributor to environmental impacts in the Alpine region.[2]

Due to the small share of settlement area in the Alpine region – in Tyrol for example only 13% – the density of the settlements in some Alpine valleys is extremely high and reaches the levels of urban agglomerations. Usable space is scarce and has to be shared between the different kinds of land-use needs. Transport infrastructure takes up a relatively high share of land, and due to the narrowness of the valleys traffic axes and residential areas are necessarily very close to each other. Furthermore, the trans-Alpine traffic flows are concentrated on only a few routes causing severe impacts in the valleys. This is aggravated by the fact that the Alpine region is situated between two main economic centres in Europe which create a high traffic demand. Natural barriers, however, prevent the population escaping from the sources of pollution or noise.

Taking into account this specific topographic and geographic situation, it can be seen that the Alpine region as a whole can be identified as a very sensitive area. This sensitiveness on the one hand and the high environmental loads and health risks on the other hand, make the reduction of pollution and noise rather urgent. A major concern in this respect is the big contribution of the transport sector to pollution, noise and land use. In order to protect the Alpine ecosystem and prevent a further increase of health risks the Alpine countries decided to establish the Convention on the Protection of the Alps (Alpine Convention, signed on 7th November 1991; it came into force on 9th March 1995).[3] This convention highlights the need for national and multilateral activities aiming at the protection of the sensitive Alpine region. As regards transport, the Alpine Convention calls for: 'transport – [with] the objective . . . to reduce the volume and dangers of inter-Alpine and trans-Alpine traffic to a level which is not harmful to humans, animals and plants and their habitats, by switching more traffic, in particular freight traffic, to the railways, in particular by providing appropriate infrastructure and incentives complying with market principles, without discrimination on grounds of nationality.'[4]

This is the Alpine topographical and socioeconomic background we start from, with the resulting basic political objective for changes in the transport sector, as manifested in the Alpine Convention, for example. The aim of the present chapter is to gain insight into present and future transport developments in the Alpine area and their economic and social implications. To this end, the results of a project called 'Environmentally Sustainable Transport' on behalf of the OECD Environment Directorate are presented. In section 2, the general set-up of this study is presented in terms of project scope and objectives. Section 3 gives definitions on pollution and noise targets for sustainable transport that need to be addressed in the long run (up to 2030). Different scenarios are described in section 4: one of these scenarios presents transportation trends and three others describe possible ways to escape current transport patterns and move towards a sustainable transport system. In section 5, conclusions are drawn as to what extent the scenarios are compatible with sustainability. Economic and social impacts are addressed in section 6, as is the policy mix that has to be implemented to redirect present transport developments. A final section concludes the chapter.

2 THE OECD STUDY ON ENVIRONMENTALLY SUSTAINABLE TRANSPORT IN THE ALPS

Looking at the severe environmental and health impacts of transport in such a sensitive area like the Alpine region, it becomes evident that specific measures to reduce these negative effects of transport are urgently needed. As many of those problems are caused by international transport flows, it is also obvious that measures on the national level are not sufficient and trans-national measures and co-operation are necessary.

To stimulate sustainable development in Alpine transport, Austria, France, Switzerland and Italy decided to work closely together within the framework of the OECD and to launch a joint pilot project on 'Environmentally Sustainable Transport in the Alpine Region' In the following, the main results of this study are reported.[5]

The aim of the present and the following sections – as the OECD study – is to define the objectives and criteria for sustainable transport in the Alps and to investigate the strategies and measures needed to attain sustainable trans-Alpine and inner-Alpine transport, at least in the four countries involved.[6]

In this study, the definition of the Alpine region is a mix of the definition set by the Alpine Convention and country-based definitions: the area according to the Alpine Convention for Austria, the whole country for Switzerland, the Alpine Convention area plus the Saône-Rhône corridor and the East Mediterranean border for France and, for Italy, the Alpine Convention plus the province of Trieste.

More specifically, the core scope and objectives of the EST project are:

- to provide an outlook for the period up to 2030 on transport and its impacts on the environment, by following the current trends in a business as usual approach;
- to determine the key features of sustainable development in transport in the Alpine region;
- to develop and model environmentally sustainable transport (EST) scenarios using the key indicators, principles and approaches elaborated on within the OECD Working Group on Transport.

This chapter reports on the stepwise approach that was taken. As a basic starting point, the criteria for environmentally sustainable transport in the Alpine region were defined. Secondly, the Alpine EST pilot study concentrated on the construction of four scenarios for the time horizon up to 2030 (see section 4). To this end, a so-called BAU scenario (business as usual) and three different scenarios for EST were carried out. While the BAU scenario assumes no major changes of policies or economic determinants compared to the past, the three EST scenarios use different approaches to describe how a sustainable transport system might be reached. Scenario EST1 focuses on technological measures while keeping transport activity at the BAU levels. In EST2, technology was set at the BAU levels and sustainability goals were realised mainly by transport demand management and modal shifts. Finally, EST3 has a wider scope and considers a combination of all relevant measures and instruments. EST3 was agreed to be the most achievable (and realistic) scenario, both in terms of balance and in terms of economic costs. Therefore, further analysis concentrated on EST3 relative to the BAU scenario.

In the following, the set of policy instruments relevant to achieving the targets of EST3 were identified and specified. The environmental, economic and social impacts of these instruments were then assessed. For the economic assessment, a distinction was made between the medium term (2015), where quantitative empirical analysis is considered more reliable, and the longer term (2030), where qualitative assessment is considered suitable. Additionally, daily life, working and transport conditions in the world of EST3 were described in a series of illustrative short 'stories'.

To ensure that mobility in the next century does not endanger human welfare, governments must go beyond conventional approaches and introduce a new generation of integrated policy and technology solutions. As stressed in UNCED's Agenda 21, the key word for future transport development must be *sustainability*. Within the OECD the joint Austrian–French–Swiss–Italian study on EST in the Alps wanted to underpin this approach and contribute to the elaboration of a sustainable transport strategy in the Alpine region.

3 THE ENVIRONMENTAL AND NOISE TARGETS IN ALPINE TRANSPORTATION FOR 2030

Environmentally sustainable transport is the application of sustainable development to the transport sector. The concept of sustainable development was first proposed by the Brundtland Commission in its 1987 report 'Our Common Future' and is defined as 'development that meets the needs of the present without compromising the ability of future generations to meet their own needs'.[7] By analogy, sustainable transport (or sustainable mobility) is transport that meets the needs of the present without compromising the ability of future generations to meet their own needs.

A sustainable transport system is one where (i) generally accepted objectives for health and environmental quality (e.g. such as those put forward by the World Health Organisation concerning air pollutants and noise) are met, as are needs for access to people, goods and services, (ii) where ecosystem integrity is not significantly threatened and (iii) where potentially adverse global phenomena such as climate change and stratospheric ozone depletion are not aggravated.[8]

In this context, a concrete definition of 'Environmentally Sustainable Transport' was developed by the OECD EST-project. What is worth noting in this context is that the basic requirement turned out to be the *needs for access* rather than 'transport' or 'mobility'. In other words, peoples' basic needs centre on the ability to reach a certain destination rather than on the desire to overcome a certain distance.

In the above attempt to define sustainable transport, common criteria for CO_2, NO_x and VOC were determined:

- The NO_x and VOC criteria are based on the need to attain the concentration levels meeting the guidelines of the World Health Organisation for air quality and eco-toxicity. Accordingly, it was concluded that NO_x and VOC emissions from transportation in 2030 should not exceed 10% of such emissions in 1990.
- The CO_2 criteria is based upon the need to prevent global warming as defined by the Intergovernmental Panel on Climate Change (IPCC). CO_2 emissions have to be reduced so that atmospheric concentration of CO_2 is stabilised at or below its 1990 level. The finally chosen reduction rates of 80% take into consideration that the relative reductions of CO_2 emissions required in OECD countries ought to be higher than in non-OECD countries in order to allow the latter opportunities for further economic development. Considering that the currently discussed Kyoto emission reduction objectives are only effectively some 2.5% (see Chapter 5) aggregated across all sectors and countries and that the emission levels in almost all industrial countries are still growing, the emission reduction chosen here is clearly very ambitious.

Moreover, definite targets for PM_{10} (particulate matter with a diameter less than 10 μm), noise and land use had to be established specifically for the Alpine region – as well as for each of the other regions and countries considered within the larger OECD-EST effort:

- Since current WHO Air Quality Guidelines indicate that no safe threshold values for PM can be set, PM emissions should be reduced to the lowest possible level. Due to lack of monitoring and research the problem of fine and ultra fine particulates in particular as regards their number and composition, which are highly relevant for their health impacts, could not be sufficiently addressed within this study. Therefore, similar to NO_x and VOC only a preliminary target for PM_{10} could be set, i.e. the level for PM_{10} in 2030 should at least not exceed 10% of such emissions in 1990.
- The noise level attained by 2030 should remain below outdoor levels that present a health concern or a serious nuisance. Due to lack of exposure data, noise could only be treated on a more qualitative basis and by illustrative examples. Nevertheless, because of its particular importance in the Alpine region due to the topographical particularities of the valleys, quantitative goals can still be set according to present legislation and the WHO threshold levels: i.e. noise levels in residential areas in 2030 should not exceed 55 dB(A) during the day and 45 dB(A) at night in 2030.

  These noise threshold levels are likely to be a fairly stringent target bearing in mind that today e.g. some 20 to 25% of the population in Switzerland are exposed to daytime road noise levels above 60 dB(A). While the level of noise emissions holds in principle for any area (mountainous or not), in the Alpine valleys the noise problem is of even greater importance, largely since the valleys are narrow and reflections from the slopes aggravate noise. This means that noise impacts are particularly high at upper slopes (mainly reserved for recreation). This constellation is worsened by the lack of settlement area compared to that available in flatter regions, and thus limits escape options.
- Land use for the movement, maintenance and storage of motor vehicles in 2030 is set so as to avoid harm to ecosystems, both directly and indirectly. Due to gaps in the data and methodological problems, no quantitative objectives could be set. Further investigations of this issue, which is of high relevance in the Alpine region, still need to be undertaken.

4 FOUR SCENARIOS FOR TRANSPORT AND ENVIRONMENTAL IMPACTS

The Business-As-Usual Scenario (BAU)

The *business as usual (BAU) scenario* was designed to capture two ideas at the same time:

- To show how a continuation of past transport trends would affect the transport system and its impacts on energy and environment by 2030.
- To state the current experts' ideas and perspectives about the future development of the transport system and its impacts on energy and environment up to 2030.

Most assumptions used to specify BAU are taken from existing regulations, recent studies, official forecasts and development programmes. This applies directly to the assumptions related to the emission factors, transport infrastructure and supply, and transport equipment ratios, for the four countries involved in the Alpine study.

It is assumed, as a general rule, that transport infrastructures and supply will develop so as to meet the future traffic requirement in the BAU. Household access to cars and motorcycles is assumed to continue to develop similarly to the past, up to saturation levels, and to stabilise afterwards. Also the demand for vehicle accessories (such as air conditioning) is assumed to increase.

Three Environmentally Sustainable Transport Scenarios (EST Scenarios)

In order to be compatible with the EST environmental quality targets, emissions will have to be reduced drastically compared to the BAU scenario. The three EST scenarios illustrate three different approaches on how this goal could be achieved:

- technological changes (EST1),
- managing and reducing the transport demand (EST2),
- a combination of technological change and demand management (EST3).

The High Technology Scenario (EST1)

The main purpose of the *high technology scenario (EST1)* is to investigate how EST could be attained by means of reducing transport's environmental impacts through technological measures alone. The scenario uses the same assumptions about the socioeconomic context of transport and the development of transport volumes as the *business as usual* trend projections.

The technology-based EST efforts focus on three key categories: (i) technological changes in road vehicle design and equipment, using best available technology; (ii) a shift from the conventional motor vehicle technology assumed in the trend projections towards new technologies and cleaner fuels with very low sulphur content (10 ppm); and (iii) broad use of alternative fuels like liquid petroleum gas, compressed natural gas and hydrogen.

The rail system also has to be improved technologically by implementing energy recuperation, improved aerodynamics and shunting technologies and more efficient rail management systems. On the input side, power plants (for traction electricity) will become more energy efficient, thus reducing CO_2 emissions. The electricity generation mix will incorporate large shares of wind-power plants and improved hydro-power plants. Ships will be equipped with fuel cells using hydrogen, which have to be produced in an environmentally sustainable manner.

The Travel Demand Management Scenario (EST2)

The *travel demand management scenario (EST2)* requires the attainment of EST solely by means of optimising mobility management, by focusing on changes in demand and supply. This involves two main factors:

- a significant reduction of transport demand for motorised traffic;
- a substantial modal shift to the most environmentally friendly transport modes (i.e. public transport, rail-freight).

The main objective for passenger transport is the reduction of transport demand for motorised transport in daily life through changes in land use and mobility patterns. To cover the remaining passenger mobility, non-motorised modes (walking, cycling) and a significantly improved public transport system will be used. An increase in the occupancy rate of the vehicles is also necessary.

Additionally, a decrease in the demand for freight transport is required to meet the reduction criteria, including changes in regional organisation (increased use of local products) and in production and consumption patterns, and changes in freight logistics and industrial location policy (e.g. fostering production clusters to reduce long-distance transport, and demanding rail access for transport-intensive production sites).

The remaining freight transport volumes must shift towards environmentally friendly transport modes and become more environmentally efficient.

Concerning the technologies of the transport means, there is no difference between the EST2 scenario and the BAU scenario. Therefore the same emission factors can be applied in the BAU and the EST2 scenarios.

The Combination Scenario (EST3)

The *combination scenario (EST3)* attempts to describe one feasible way to achieve EST goals in the same general socioeconomic context as that of the BAU scenario. It combines the technological approach of the EST1 scenario and the transport demand approach considered in the EST2.

The EST3 has been designed with the idea of balancing the long lead times in the commercial development of entirely new technologies – which from an economic and social viewpoint is more easily acceptable – with the required transformation of the existing transport system. In the EST3, the progressive internalisation of environmental externalities in the price of transport services contributes to an increase in the overall and sectoral transport costs, making advanced technologies, whose elaboration and diffusion is supported by government policies, progressively more attractive for producers and consumers.
The main strategies in the EST3 scenario are:

- decoupling economic growth from transport and related environmental impacts;
- a reduction of transport demand by changes in land use and mobility patterns as well as in production and consumption patterns;
- a more efficient use of vehicles and infrastructure, as well as broader use of information technologies;
- a significant shift of passenger transport towards non-motorised transport, rail and public transport, and freight transport towards rail, inland and coastal shipping and combined transport;
- a considerable improvement of fuel quality and the technology of road vehicles towards ultra low emission vehicles and partly towards zero emission vehicles based on sustainably produced hydrogen fuel cells;
- and a positive development in rail technology and rail management and logistics towards higher efficiency and an improvement of power plants for rail electricity generation.

5 CONCLUSIONS FROM THE SCENARIO ANALYSIS

The BAU Diverts The Transport System Away From Environmental Sustainability

The BAU scenario shows very clearly the impact of unsustainable development of transport in the Alpine region with respect to the goals defined in section 3. Freight transport would grow by 76%, whereby trans-Alpine traffic (traffic crossing the Alps) alone would multiply almost by a factor of 3 (see Figure 12.1a). Passenger traffic would grow

(a) Transport volumes (freight and passenger)

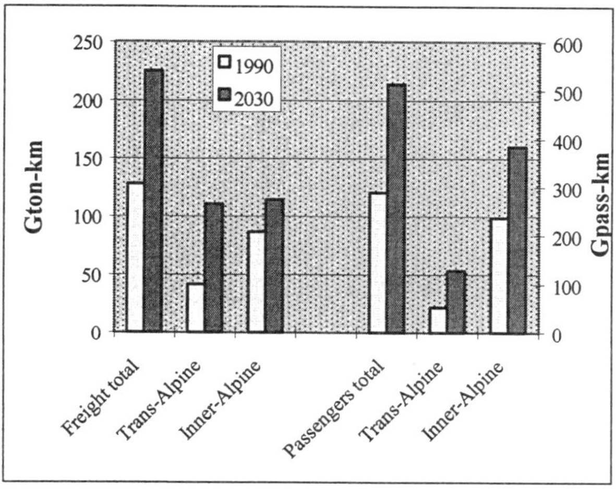

(b) Emissions from transport

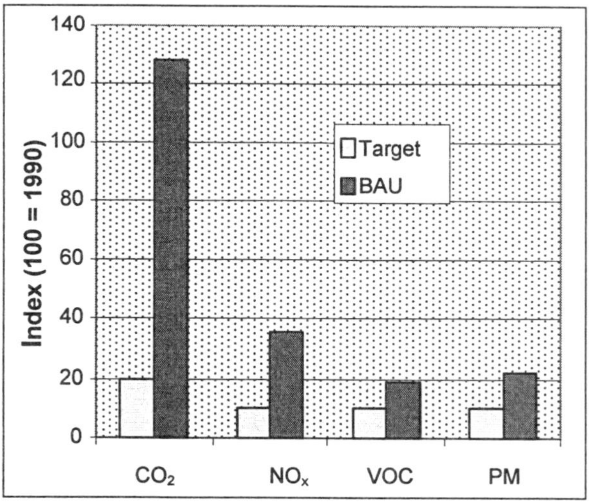

Figure 12.1: The BAU scenario traffic and emissions in 2030 against the EST goals, Alpine region

by 76%, with trans-Alpine traffic alone increasing by a factor of 2.4. As Figure 12.2 indicates, road would capture more than 90% of the passenger traffic increase, and nearly 80% of the freight traffic increase. Finally, Figure 12.1b translates traffic volumes into emissions:

(a) Freight transport – modal split

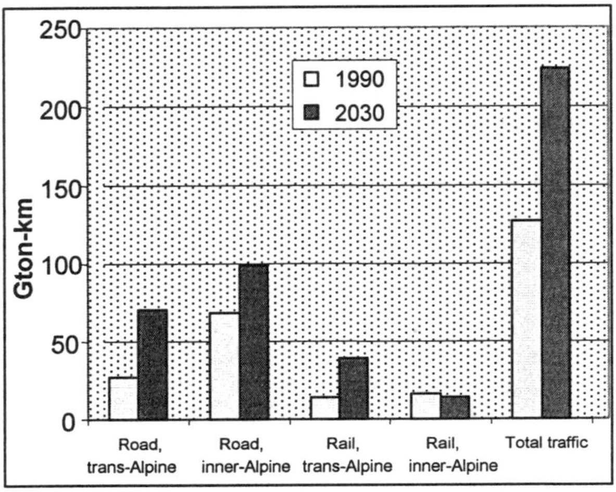

(b) Passenger transport – modal split

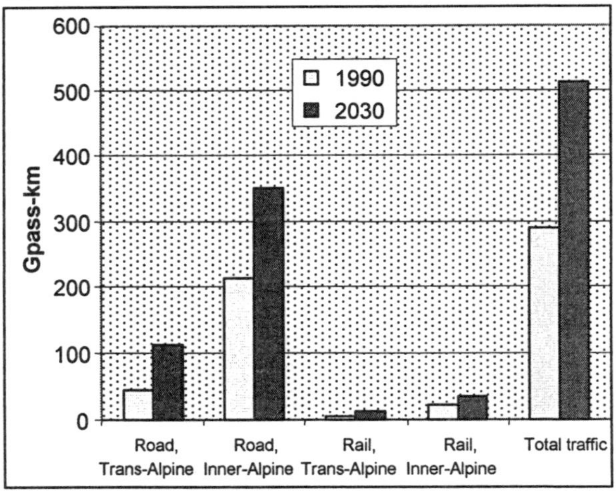

Figure 12.2: Modal split for the BAU and EST goals in 2030, Alpine region

CO_2 emissions would increase by 30% compared to 1990 and emissions of NO_x, VOC and PM would attain levels twice or three times bigger than needed for sustainability (as defined in section 3), despite the fact that important technical improvements are assumed in BAU (e.g. particle filter and De-NO_x-catalytic converters for diesel engines, fuel-efficient cars with average consumption below 5 l/100 km).

Obviously, the development of road and rail infrastructures already planned and necessary to accommodate the traffic development in BAU up to 2030, as well as the indirect impacts on land use of infrastructure and traffic development, would dramatically increase the share of land devoted to transport activities, in particular in that most sensitive part of the Alpine region: the valleys hosting the international routes. Despite the fact that technological improvements have been assumed, the situation would strongly deteriorate both in terms of land use and noise in BAU as compared to the present situation and would move certainly away from EST goals.

Sensitivity analysis carried out on the transport demand elasticities shows that trans-Alpine freight traffic might develop more slowly than assumed in the BAU. But even in this case, the BAU would still be far away from what would be required for sustainability.

Scenario EST1: Technological Breakthrough is Necessary but not Sufficient

From a pure transport point of view, the sustainability criteria for emissions from pollutants and, partially, noise could be met by a technology breakthrough. As shown by the EST1 results, adequate technologies are either already known or in their experimental phase, but their development and market penetration would need to be much faster than in the past and also costs might be quite high.

Three major problems remain:

- The CO_2 emission sustainability criterion appears to be the most difficult to achieve through technological development alone. It supposes a large contribution of on-board hydrogen fuel cells that have to be produced CO_2-free and distributed on a large scale.
- The development of traffic considered in the BAU would increase the need for new road and rail infrastructures beyond what is already planned, and would make the direct and indirect impacts on land use much bigger than today, especially in the Alpine valleys. Technology alone would not solve this problem and is therefore unlikely to contribute to sustainability from this viewpoint.

- Technological improvements would significantly reduce noise emissions at source (road vehicles or trains), but probably not sufficiently to reach EST goals. Additional noise abatement walls and tunnels would certainly be required close to living areas.

Scenario EST2: Transport Demand and Supply Management is also Necessary but not Sufficient

From a purely transportation system viewpoint, sustainability criteria for pollutants emissions could be reached by solely changing the transport demand and supply pattern, with no additional technology compared to the BAU: as shown by the EST2 results, internalising environmental externalities in transport prices, promoting the supply of environmentally friendly transport modes, giving back the cities to pedestrians, bicycles and public transport, might result in pollutants emissions compatible with sustainability criteria.

Such an evolution of the transport demand and modal allocation would require intensive structural changes both in the transport sector and in production and distribution patterns. Such structural changes have to be trans-national, implying the need for extensive international co-operation.

A necessary prerequisite for the EST2 would therefore be a tremendous change in behaviours, mentalities and social priorities. Moreover, it would also be necessary to foster in-depth transformation of the urbanisation schemes towards greater but more sustainable density, in particular in those big conurbations with no shortage of space, as is often the case in France. Daily trip distances would have to be massively reduced to walking/bicycling distances and living/working/shopping areas would have to concentrate along public transport routes, which is in fact the opposite of current trends.

Again, the CO_2 emission sustainability criteria appears to be the most difficult to achieve through changes in transport demand and supply patterns. In particular it requires a dramatic shift from road to rail for both freight and passengers, and a complete CO_2-free electrification of all additional rail traffic from today on.

Scenario EST3: Transport Sustainability can be Achieved through the Combination of Technological Breakthrough and Transport Management Changes

Several combinations of technological breakthrough and transport management changes may help attain transport sustainability in 2030, fulfilling the EST criteria for pollutant emissions and, partially, for noise. The increases of total freight and passenger traffic would be slightly lower than in the BAU (around 55% for both between 1990 and 2030), but the road freight traffic would increase up to a turning point and decrease to stabilise in 2030 to its 1990 level, and the road passenger traffic in 2030 would be only 20% above its level of 1990 (see Figure 12.3). Total trans-Alpine road traffic would also peak and then decrease down to its 1990 level. These changes however would require concerted trans-national actions, in particular with regard to trans-Alpine transport demand, and would also be required in countries outside the Alpine region. Purely local measures would certainly not be sufficient to reduce the environmental loads to a level of sustainability, since they have limited effects on technology and on trans-Alpine transport demand.

In addition, the problem mentioned above about pollutant emissions from electricity generation still remains, in particular in the case of a large development of H_2 fuel cells, if electricity is not generated with CO_2-free techniques. The sensitivity analysis shows that sustainability can still be achieved with only liquid petroleum gas or methanol fuel cells, if electricity generation for supplying railways is massively CO_2-free. This seems to be more easily feasible since a large part of electricity requirements for railways is off-peak and more suitable for non-fossil fuel generation.

Extensive development of rail infrastructures would be needed in any combination of technological breakthrough and transport management changes in order to fulfil sustainability criteria for pollutant emissions. This may raise questions with respect to sustainability criteria for land use, unless major changes occur in the management of rail infrastructures and new priorities are set for the future development of new infrastructures in order to limit the development of road infrastructure.

EST goals for noise might be difficult to reach, despite technological improvements and related reductions in noise emission at source, even for rail rolling stock. This results from increasing rail and remaining road traffic, which could only be offset by additional noise abatement measures.

In any combination of technological breakthrough and transport management changes such as to meet sustainability criteria for pollutants, significant changes in behaviours, mentalities and social priorities would be required. In particular, transformation of the urbanisation schemes, specifically, from sprawl to more mixed land-use planning, would be needed.

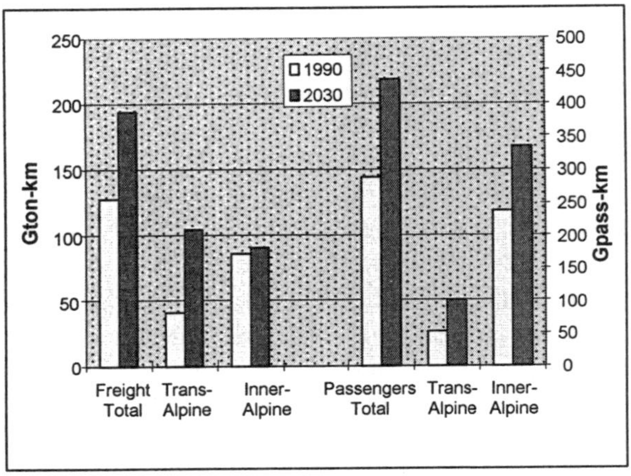

Figure 12.3: Freight and passenger transport in the EST3 Scenario (1990–2030)

Detailed Quantification of Transport Volume and Emission Impact for EST3

Table 12.1 gives the detailed split-up per country of transport volumes for passenger and freight transport in the EST3 scenario.

Table 12.1: Traffic development in the Alpine region, 1990 and 2030, EST3 scenario

EST 3	1990					2030				
	A	F	CH	I	Total	A	F	CH	I	Total
Road										
*Freight (Gtkm)**	9.1	60.2	10.5	15.2	95.0	9.8	51.9	15.9	13.3	90.9
Trans-Alpine	5.8	13.0	1.1	6.7	26.6	5.8	14.7	1.4	8.0	29.9
Inner-Alpine	3.3	47.2	9.4	8.5	68.4	4.0	37.2	14.5	5.3	61.0
Passengers (Gpkm)	31.6	119.6	81.1	28.3	260.6	30.4	165.5	93.6	23.5	313.0
Trans-Alpine	3.0	35.5	5.4	1.4***	45.3	2.7	36.5	7.4	1.3	47.9
Inner-Alpine	28.6	84.1	75.7	26.9	215.3	27.7	129.0	86.2	22.3	265.0
Total (Gvkm)	19.3	80.6	50.6	17.0	167.5	17.7	91.0	48.0	9.2	165.9
Trans-Alpine	1.8	15.9	2.5	1.3	21.5	1.6	15.0	3.1	1.0	20.7
Inner-Alpine	17.5	64.7	48.1	15.7	145.9	16.1	76.0	44.9	8.2	145.3

Table 12.1 (cont.)

EST 3	1990					2030				
	A	F	CH	I	Total	A	F	CH	I	Total
Rail										
Freight (Gtkm)	7.1	12.7	8.9	3.1	31.8	26.4	40.0	23.7	15.5	105.6
Trans-Alpine	4.7	2.8	4.8	2.8	15.0	23.0	27.1	16.3	11.9	78.3
Inner-Alpine	2.4	9.9	4.1	0.3	16.7	3.4	12.9	7.4	3.6	27.2
Passengers (Gpkm)	4.7	6.7	15.2	2.5	29.1	13.1	65.0	31.4	14.0	123.5
Trans-Alpine	**	4.7	1.8	0.5	6.9	**	46.0	5.0	1.7	52.7
Inner-Alpine	4.7	2.0	13.4	2.0	22.2	13.1	19.0	26.4	12.3	70.8
Water (Gtkm)										
Inner-Alpine freight	0.0	1.5	0.0	0.0	1.5	0.0	2.2	0.0		2.2
Total Traffic										
Freight (Gtkm)	16.2	74.4	19.4	18.3	128.2	36.2	94.1	39.6	28.8	198.8
Trans-Alpine	10.5	15.8	5.9	9.5	41.7	28.8	41.8	17.7	20.0	108.3
Inner-Alpine	5.7	58.6	13.5	8.8	86.5	7.4	52.3	21.9	8.9	90.5
Passengers (Gpkm)	36.4	126.3	96.2	30.8	289.7	43.5	230.5	125.0	37.5	436.5
Trans-Alpine	3.0	40.1	7.1	1.9	52.2	2.7	82.5	12.4	3.0	100.7
Inner-Alpine	33.3	86.2	89.1	28.9	237.5	40.8	148.0	112.6	34.5	335.9

Notes:

* Gtkm: Giga-ton-kilometres;

** Included in 'inner-Alpine' traffic;

*** In Italy the pass-km of trans-Alpine road traffic are smaller than the pass-km of the other countries because of the geographic peculiarity of Italian Alpine regions, and because the main part of the traffic crosses the Alps north-south. For these reasons the average distance covered is only about 70 km.

Table 12.2 gives the resulting emissions for the EST3 scenario. The EST target as specified in section 3 above is met for CO_2, and levels for the remaining airborne emissions are much lower than target.

As noise represents a local environmental problem it does not make sense to evaluate aggregate Alpine noise numbers, but a representative analysis was carried out for one main Alpine route, the Brenner route.

Based on the situation in 1990 and the general results of the BAU and EST3 scenarios, the daily number of vehicles and trains along the Brenner route was estimated (see Table 12.3).

Table 12.2: Emissions in the Alpine region, 1990 and 2030, EST3 scenario

EST 3	1990					2030				
	A	F	CH	I	Total	A	F	CH	I	Total
CO_2 emissions (Mt/a)	6.2	22.0	13.8	3.9	45.9	1.2	4.4	2.7	0.7	9.0
trans-Alpine, freight	0.6	0.9	0.2	0.5	2.2	0.1	0.2	0.0	0.1	0.4
trans-Alpine, passengers	0.2	2.8	0.5	0.1	3.7	0.0	0.6	0.1	0.0	0.7
inner-Alpine	5.3	18.3	13.1	3.3	39.9	1.0	3.6	2.6	0.6	7.8
NO_x emissions (kt/a)	44.0	193.6	107.0	62.7	407.3	0.9	13.1	2.9	4.9	21.8
trans-Alpine, freight	6.9	10.4	2.3	12.0	31.5	0.3	0.7	0.2	0.9	2.1
trans-Alpine, passengers	1.7	31.7	4.2	2.0	39.7	0.0	0.5	0.0	0.1	0.7
inner-Alpine	35.4	151.5	100.4	48.8	336.2	0.5	11.9	2.7	3.9	19.0
VOC emissions (kt/a)	27.7	108.1	90.1	27.2	253.2	0.6	2.0	0.7	1.5	4.8
trans-Alpine, freight	0.7	1.1	0.2	2.4	4.5	0.0	0.1	0.0	0.1	0.2
trans-Alpine, passengers	1.3	20.1	2.3	0.4	24.1	0.0	0.2	0.0	0.0	0.2
inner-Alpine	25.7	86.8	87.6	24.4	224.6	0.5	1.8	0.7	1.4	4.3
PM emissions (kt/a)	1.8	9.0	2.9	2.1	15.8	0.0	0.3	0.1	0.2	0.5
trans-Alpine, freight	0.4	0.5	0.1	0.7	1.7	0.0	0.0	0.0	0.1	0.1
trans-Alpine, passengers	0.0	0.5	0.0	0.0	0.5	0.0	0.0	0.0	0.0	0.0
inner-Alpine	1.5	8.0	2.7	1.4	13.6	0.0	0.2	0.0	0.1	0.4

In the EST3 scenario, in addition to the number of freight trains, the number of intercity/interurban passenger trains will also rise substantially on the Brenner route. Most of the intercity/interurban passenger trains and the freight trains will use the Brenner base tunnel.

Along the trans-Alpine motorway A13, the daily traffic volume will be reduced from 14 100 to less than 12 000 vehicles per day, keeping the share of heavy duty vehicles (HDV) approximately constant.

Corresponding to the railway noise map, zones with noise levels above 50 dB(A) are reduced. Nevertheless it has to be stated that the zones with noise levels above 45 dB(A) expand. Thus, while areas exposed very strongly are reduced in size, those strongly exposed do increase.

Table 12.3: Traffic volume on highway A13 and railway line along the Brenner route at Matrei, 1990, BAU 2030 and EST3 2030

	1990	BAU 2030	EST3 2030		
Highway A13					
Passenger cars	9 870	17 957			8 411
HDV*	4 230	12 478			3 435
ADV**	14 100	30 435			11 846
Share of HDV	30 %	41 %			29 %
Railway line			Present track	Tunnel	Total
EC/IC trains	21	21	7	32	39
Local trains	31	31	61	0	61
Freight trains	60	152	8	249	257
Other trains	24	24	8	4	12
Total	*136*	*228*	*85*	*285*	*369*

Notes:
To suite the local situation of the Brenner route, in the EST3, for local public transport a light rail system using the existing tracks (comp. Stubaitalbahn) is assumed;
* HDV = heavy duty vehicle;
** ADT = average daily traffic.

From Figure 12.4 it can be seen that compared to the BAU scenario, the noise levels caused by rail and road traffic are lower in EST3, which is due to the shift of most of the intercity/interurban passenger trains and freight trains to the Brenner railway base tunnel and due to reducing the traffic volume on the motorway. New noise protection measures along the railway line and the motorway also contribute to the reduction. The higher number of local trains (light rail) on the existing Brenner railway line as estimated in Table 12.3 has only a small influence on the noise level. In the EST3 scenario nearly all settlement areas are no longer affected by noise levels above 50 dB(A).

Further Research Demand for Scenario Development

Due to constraints in the availability of data and modelling methodology, in the OECD study special attention was paid to EST emission criteria for PM, NO_x, VOC and CO_2. This of course does not paint the full picture of the EST. In particular in the Alpine region, land use for example appears as a crucial additional indicator of sustainability. Similarly, environmental issues related to aviation could not be sufficiently addressed and solved in the context of the study reported in this chapter, and also necessitate further in-depth investigation in the future.

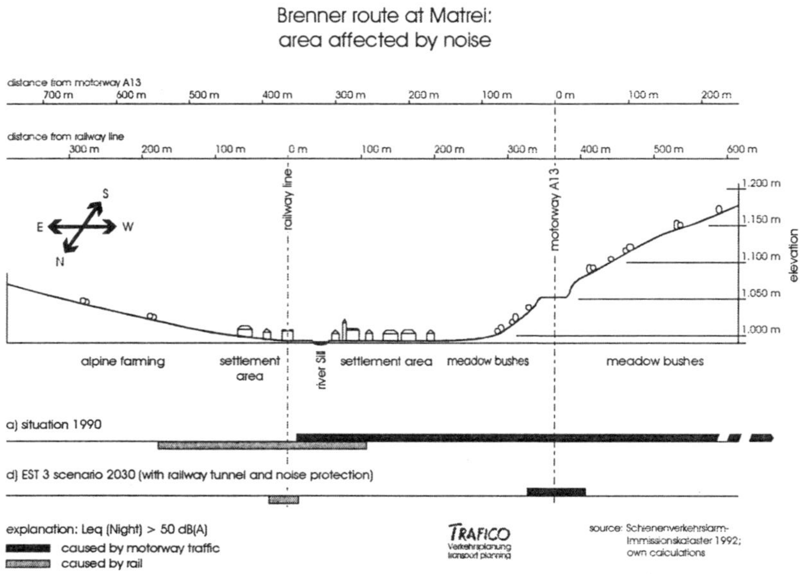

Figure 12.4: Noise impact zones during night time along the Brenner route at Matrei in the EST3 2030 compared with 1990

6 ECONOMIC AND SOCIAL IMPLICATIONS OF THE COMBINED TECHNOLOGY AND TRANSPORT MANAGEMENT SCENARIO

Balancing Efforts to Promote Changes in Technology and Transport Demand Trends

As the previous section indicated, CO_2 turned out to be the most difficult goal to achieve. In other words, when the CO_2 targets were met the other targets were well met. Therefore, the Balance Of Effort (BOE) analysis aims at assessing the relative contributions of four factors to the attainment of the CO_2 reductions required for the EST3 scenarios in the four countries. The comparisons are with the respective BAU scenario. The four factors are:

1. Reduced emissions per unit of transport activity, i.e., passenger kilometres or tonne kilometres, from same vehicle type through *technological change* or *vehicle downsizing*.

2. *Reduced transport activity*, i.e., fewer passenger kilometres or tonne kilometres.
3. Reduced emissions per unit of transport activity through use of *more efficient vehicle types*, i.e., through mode shifts.
4. Reduced emissions per unit of transport activity through using the same vehicle type more efficiently, i.e. *higher occupancy.*

The specific purpose of the BOE analysis is to attach a percentage value to each of the four factors. According to this methodology, the results for the Alpine region are:

- Technological change and vehicle downsizing contribute to about two-thirds of the effort towards achieving the difference in CO_2 emissions between the EST3 and BAU conditions.
- In the remaining one-third of the effort, about 40% comes from reducing overall transport activity, 35% from shifting to more fuel-efficient transport modes and 25% from using vehicles more efficiently.

The country-specific BOE results for the four Alpine countries are displayed in Table 12.4.

Table 12.4: Balance of effort for Austria, France, Switzerland and Italy

	Estimated share of effort assigned to each method			
Type of emission reduction per unit of activity	Austria	France	Switzerland	Italy
(1) Technological change or vehicle downsizing for each vehicle type	61%	66%	67%	42%
(2) Reduction in overall passenger kilometres or tonne kilometres	17%	13%	10%	13%
(3) Use of more efficient vehicle types, i.e. through modal shifts	16%	11%	8%	27%
(4) More efficient use of each vehicle type, i.e. higher occupancy rates	6%	11%	16%	18%

Instruments and Policy Measures to Reduce the Environmental Impact of Alpine Transport

Most instruments and policy measures necessary to reach EST3 targets are already in place in the Alpine countries, or on their political agenda. Thus, the challenge is mainly in terms of stringency and timing, or phasing of instruments. Considering the substantial barriers faced during previous attempts to direct environmental policy in the past, this is still likely to be a very difficult undertaking. The suggested instruments include the following:

- economic instruments for fair and efficient pricing and taxing in order to internalise the external costs of CO_2 emissions and pollutants, and the use of public space and noise;
- regulatory instruments aiming at fixing 'the rules of the game' for the fuel specifications, the emissions levels of the vehicles for a series of pollutants and noise, and for the use of public space and infrastructure;
- financial instruments to address the funding requirements for the development of infrastructure and services of environmentally friendly modes such as rail, public transportation, bicycles and walking, and for research, development and demonstration of new transport technologies (new energy forms, new vehicles, new modes);
- institutional instruments for the promotion of change in the governance arrangements to aid implementation of trans-national and national transport demand management programmes and regulations, and for the integration of urban and regional planning and environmental objectives and requirements in transport policies and programmes;
- educational and hortatory instruments to raise the public awareness about the environmental damage of transport activities, and to raise interest in pilot actions and research for innovative sustainable transport solutions.

Most country research teams put a particular emphasis on economic instruments. Transport cost shares currently financed directly by charging transport are extremely low (particularly so for road transport). For example, in Austria road car transport funds only 38% of the full costs it causes (including accident and environmental cost, but not congestion), road freight transport only finances 20% (Axhausen et al. 2002). Thus there is a strong argument for more efficient (road) transport pricing. In order to achieve the substantial CO_2 emission reductions of the EST scenarios, relying on road pricing alone may be not feasible. For the final emission reduction stages up to 2030 further net charging of the transport sector may have to be avoided in order to ease implementation. This could be achieved by the introduction of a CO_2 tradable permit system and an initial distribution based on some grandfathering system. While the net costs for the transport sector (or the economy as a whole if permits are tradable economywide) would thus remain stable, the incentive to reduce emissions would remain, and be strongest for those emitters that could do so most cheaply.

When the above instruments are specified and prioritised for the various Alpine study countries, it appears that for the most part, these instruments and policy measures correspond to decisions already taken or in the 'pipeline'. The essential difference here, compared to what is required to reach the EST3 targets, is far more the necessary magnitude of the policy measures expected during implementation than the specific nature of the policy measures themselves. This is an important point, since it indicates that the underlying political and social acceptance or desirability of these types of instruments is to a large extent already present. Yet to attain the necessary level in stringency is a difficult task.

A transition to an environmentally sustainable transport system would require action from several different parts of society on an international, national, regional and local level, including politicians, authorities, scientists, NGOs, the private sector and individual citizens. Implementation of a successful, environmentally sound transport policy could consist of three phases:

1. *Acceptance phase* Discussion of criteria, goals and measures in order to attain the EST. Development and adaptation of strategy and argumentation. Information on long-term goals, health and environmental effects of transport and their economic implications is important. Additionally, instruments addressing behaviour and motivating people to promote sustainable transport.
2. *Adaptation phase* Adjustment of current policies towards a more environmentally friendly transport policy. Stepwise implementation of various instruments, such as economic instruments, standards and regulations, or a change in zoning philosophy. Consideration of the interdependence of different measures and instruments.
3. *Implementation phase* Integration of the principles and goals of sustainable development into transport and transport-relevant sectors.

In the particular case of the Alpine region, this process must also take into consideration the issue of convergence of the strategies, instruments and policy measures among the four Alpine countries. Also, part of the instruments and policy measures are to be implemented at the international level, in particular that of the EU, and the implementation of these does not depend so much on the national political agenda, but on that of the international institutions involved.

Finally, there is an obvious interlinkage with policy directed at sectors other than transport. The more production structures are integrated, the less transport is demanded. Policy that fosters such locally vertically integrated production or that fosters consumption of locally produced goods (take organic food for example) at the same time reduces transport demands.

Economic Implications

Research in the four countries chose different approaches to assess and quantify the economic implications of the EST3. Nevertheless the final results of the individual research efforts were rather close again. We therefore shall report here on only one country in more detail, Austria, giving a representative example.

The instrument package to achieve the EST objectives in Austria includes a development of road pricing towards full cost internalisation by 2015 and a complementary CO_2 permit trading scheme starting in 2015, for both freight and passenger transport, as the two central elements. A range of further instruments has to ensure the achievement of specific objectives with respect to local impacts and noise impacts.

As the acceptance of methods for social and economic *quantitative* assessment of such distant scenarios as the full EST scenario is limited, the OECD EST group agreed that quantification of impacts should extend only until 2015 or 2020, while qualitative methods are to be used for the years thereafter. For Austria the time span to 2015 has been chosen for quantification, as this year marks an important milestone of the EST instrument package: full cost internalisation for noise, NO_x, VOC and other air pollutants, and partial internalisation for CO_2. The quantitative impacts of the EST instrument package implementation have been discussed, and the qualitative impacts have been developed and discussed in a series of expert workshops.

The quantification of the economic implications of the EST freight transport instrument package has been accomplished by generating and applying a computable general equilibrium model, as this approach allows for price-dependent endogenous adjustment of the input coefficients. By an elaborate matching of a 32-sector economic database with the ten-sector transport database, the categories of domestic long-distance transport, export and import transport and transit transport can be integrated in an economic model of Austria in terms of ton-kilometre (tkm) levels of freight transport per sector.[9]

The quantitative analysis is based on the major fractions of freight transport. It compares the BAU scenario on the one hand, and the EST3 scenario on the other hand. The analysis reveals the following:

- Overall economic impacts induced by the EST instrument package are rather small. GDP growth is lower by less than 0.1% p.a. up to 2015 in the EST3 case. This is in stark contrast to the substantial emission reductions achieved.
- The effects of the EST3 instrument package in the labour market are beneficial. Unemployment under EST is roughly 0.5% lower. The rise in employment in the EST3 scenario amounts to roughly 20 000 employees by 2015. This rise in employment is mainly due to the nominal wage rate adjusting with a slight time lag to the price rise triggered by road pricing (this transfers the experience gained from both oil price shocks to what would be found under a significant transport cost price rise).
- The sectoral shifts induced by the EST3 instrument package indicate a slightly higher growth in agriculture, roughly equal output levels in most manufacturing sectors and lower gross production in public services.

Road and rail pricing at the cost internalisation levels given for 2015, if implemented as the *only* instruments, are likely to achieve the EST3 transport volume objectives. In rail transport the EST3 ton-kilometres level objective in this case would be more than fulfilled. Thus, the further instruments of the EST3 instrument package are necessary to ensure:

- regional meeting of the objectives;
- meeting of the noise objective;
- meeting the objective in short-distance transport;
- lowering of the overall transport reduction burden by a net (shadow cost) subsidy to rail transport (i.e. infrastructure), increasing rail transport.

The sensitivity of the quantitative results with respect to model parameters is very small for overall economic indicators. Modal transport volumes, however, react significantly to parameter change.

Qualitative analysis of both freight transport beyond 2015 and passenger transport reveals for the EST3 scenario, relative to the BAU scenario, that:

- overall welfare (material welfare plus benefits from environmental and noise nuisance improvements) is higher for the EST3 scenario in the long run, while there may be adjustment difficulties in the transition to EST;
- employment increases (at least in the transport sector) due to the rising service intensity and transport logistics;
- in the transport sector we observe a range of trends, including more life-cycle programmes in the automobile industry and a decrease in average vehicle turnover times due to more stringent standards;
- there is relatively strong mutual reinforcement between the achievement of sustainability objectives in the transport sector, sectors of organic farming and forestry and environmental technologies.

Finally, note that effects of the policy mix are not always unambiguously positive with respect to the sustainability objective: Although designed according to these objectives, some instruments may have both positive effects (e.g. increased efficiency), and negative effects (e.g. increase in transport demand); for example, subsidising renewables or liquid petroleum gas is relevant to promote cleaner fuels, but may slow down the process of shifting from cars to public transportation.

Social Implications

Daily life in the EST3 is close to what people would be ready to accept today, rather than being a revolution. To make the EST more visual and understandable, images of future EST life were produced telling the daily life story of the 'Alpinetree' family, or that of their relatives or neighbours in the EST3. To explain how things are produced, transported and distributed in this world, how people use their cars, etc., there was no need to invent futuristic or exotic events. In fact, it is just a matter of transposing the various experiments and situations that already exist in different countries from one place to another. Many of these policies have been in operation for quite some time and have found ready acceptance among a substantial section of the public because they have often led to a considerable increase in the quality of life, e.g. car-free urban centres, bicycle lanes, modern trams and friendly housing compounds.

Although a fiction, the story, based on all technical and organisational assumptions of the EST3, does not show a world very different from today. One might probably say it is a little more attractive. People still move freely, albeit less by car; the car industry has not grown, but is specialised in high-tech cars that it exports all over the world; rail and public transport have become a new 'locomotive' of the economy, supporting sectors such as construction, equipment material and design; restoration, maintenance and repair of cars are not so predominant as they used to be, and many new services have emerged for public transportation, logistics, travel assistance, car rental and leasing.

This fiction and the EST3 scenario in general confirm the finding that increased 'mobility' is liberating and socially progressive up to a certain point. Beyond this point it may become socially destructive – especially when accompanied by increasing disparities in levels of mobility. Several examples underline this phenomenon:

- Access, measured in the amount of travel required to reach people, goods and services, may decline due to car dependency. The process of increasing car dependency is one of increasing segregation and polarisation. All those without a driving licence, but otherwise capable of independent mobility, face a growing threat to their independence. Children in particular have lost considerable independence as society's dependence on the car has increased.
- According to Kenworthy et al. (1997), it has been established that beyond a certain point (about 7 500 km per capita annual car travel) increased driving, and the resulting car-dependent transport system and low density land-use patterns, impose economic costs that outweigh marginal economic benefits. The study indicates that commuter travel times do not vary significantly among the 37 cities examined, that the portion of income devoted to commuting is highest in car-dependent cities (particularly lower income, developing cities are car-dependent), and that transit service efficiency is lowest and vehicle accident rates per capita are highest in car-dependent communities.
- Considering the expected development of motorisation, a very strong growth of land used for transport infrastructure has to be expected for the future. Improving the traffic system, however, leads to a more than proportional traffic growth, valid for all transport modes, so-called induced traffic, which is true for all transport modes. Historic examples show that the construction of new infrastructure often triggered off additional land use: Such induced land use takes place when land-use patterns adjust to the newly available capacity and the resulting spatial allocation of activities.

The EST3 allows such trends to be broken form a social point of view. It thus appears more attractive than a world based on the continuation of current trends.

7 CONCLUSIONS

This chapter has reported in a nutshell the main results of a common research effort by Austria, France, Italy and Switzerland within the framework of the OECD, the objective of which was to study the options and implications of reducing environmental impacts of transport in the Alpine region. This research effort was part of the so-called 'environmentally sustainable transport' (EST) project of the OECD.

The results indicate that neither a purely technological solution is feasible, as it would be too expensive, nor a pure transport demand management approach, as the latter would require shifts in demand and daily practices that might not be achievable easily within a reasonable time. However, even for very strict environmental targets, such as a reduction of CO_2 emissions from transport by 80% in 2030, relative to the 1990 level, the analysis also indicates that a combination of technological and transport management policy measures can well achieve environmentally sustainable transport in the Alpine region. Furthermore, in the four countries most of these types of policy measures are already in place (and thus accepted by type) or at least already under discussion, and so it is mainly a question of their respective stringency.

Such a significant change in transport demand and modal allocation would require intensive structural changes both in the transport sector and in production and distribution patterns. Both areas of structural change would not be restricted to a particular geographical area and therefore would need transnational co-operation.

The economic implications of choosing such a development path as judged by traditional indicators such as GDP have been found to be minor. In terms of employment, a slight rise is likely. The main economic effect is a restructuring of the economy, with short-term impacts being larger than the long-term impacts. Finally, interesting insights into a sustainable transport world and its social implications can be gained by employing so-called 'daily life scenarios'.

NOTES

1. This chapter reports on and draws on OECD (2001). Financial support for this research was supplied by the Austrian Federal Ministry for Environment, Youth and Family (Section Transport, Mobility, Regional Planning and Noise), the French Agence de l'Environnement et de la Maîtrise de l'Energie, the Italian Ministry for Environment, and the Swiss Agency for the Environment, Forests and Landscape. Direction by the project steering committee, consisting of Massimo Cozzone (Italy), Harald Jenk (Switzerland), Alain Morcheoine (France), Robert Thaler (Austria), Gloria Visconti (Italy) and Peter Wiederkehr (OECD) is thankfully acknowledged.
2. For example, for the region of Tyrol a provincial governmental document states that 'In Tyrol emissions from [road]transport are the determining factors for air quality, especially with respect to nitrogen oxides, carbon monoxide and particulate matter, partly also with respect to ozone. Concentration levels next to highways are above even those in the centre of Innsbruck.' (Amt der Tiroler Landesregierung (2000, p.28); translation by the authors). Within transport, nitrogen monoxide emissions show an almost perfect correlation with truck activity.
3. A Transport Protocol, based on the Alpine Convention, was adopted by the contracting parties in May 2000 during the 16th Meeting of the Standing Committee of the Alpine Convention. The European Commission is preparing a Council Decision on the signature of this Protocol (COM(2001) 18 final).
4. European Communities (1991), Convention on the Protection of the Alps (Alpine Convention) – English Translation, Official Journal of the European Communities, L 061, 12/03/1996, 0032-0036, Article 2(2).
5. The project documentation is available at the OECD Environment homepage http://www.oecd.org/env/ccst/est.
6. Trans-Alpine traffic is defined as traffic over the main Alps border passes in north–south direction or vice versa, e.g. traffic passing along the Brenner route. Inner-Alpine traffic is defined as all the traffic (passenger and freight) with origin or destination in the study area that does not cross a main Alpine pass and a border during the same journey.
7. World Commission on Environment and Development (1987, p.43).
8. OECD Environment Directorate (1996, p.4). (http://www.oecd.org/env/docs/estbroche.pdf)
9. The structure of the model is described in full detail in Steininger (2001, pp.113–117).

BIBLIOGRAPHY

Amt der Tiroler Landesregierung (2000), 'Verkehrsentwicklung in Tirol 1999', Innsbruck.

Axhausen, K., M. Herry, H. Link, M. Schuster, and N. Sedlacek (2002), 'Wegekostenrechnung 2000 fuer den Verkehrstraeger Strasse in Oesterreich', report for the Austrian Federal Ministry of Transport, Innovation and Technology. Vienna: Buero Herry.

Kenworthy J., F. Laube, P. Newman and P. Barter (1997), *Indicators for Transport Efficiency in 37 Global Cities*, Sustainable Transport Research Group, Murdoch University (Perth), Washington DC: World Bank.

OECD (2000), 'Est! Environmentally Sustainable Transport: Futures, Strategies, and Best Practices, Synthesis report of the OECD project on Environmentally Sustainable Transport', presented on the occasion of the International Est! Conference, 4–6 October 2000, Vienna, Austria.

OECD (2001), 'E-S-T Environmentally Sustainable Transport "Alpine Region" – Austria, France, Italy, Switzerland. Synthesis Report Phase 3', Paris: OECD.

OECD Environment Directorate (1996), 'Project on Environmentally Sustainable Transport', Brochure, Paris. (http://www.oecd.org/env/docs/estbroche.pdf)

OECD Environment Directorate (n.y.), Flyer: 'Project on Environmentally Sustainable Transport', Paris: OECD.

Robinson (1990), 'Futures under glass: a recipe for people who hate to predict', *Futures* 22(8), pp. 820–842.

Steininger, Karl W. (2001), *International Trade and Transport, Spatial Structure and Environmental Quality in a Global Economy*, Cheltenham: Edward Elgar.

World Commission on Environment and Development (1987), *Our Common Future*, Oxford: Oxford University Press.

13. Conclusions and Future Research Prospects

Karl W. Steininger and

Hannelore Weck-Hannemann

The diverse environmental changes we indicated in the introductory chapter (such as climate change, land-use change and its implications for biodiversity, water-cycle change) have been analysed from various perspectives by the contributors throughout this volume. Although they are global, their implications take specific local forms. This is true in both directions: for the impacts of global environmental change on particular regions and for the resulting feedback effect from these regions to other locations and to the overall global change phenomena. This volume has selected alpine regions for analysis, as such regions are particularly exposed to global change (e.g. see Chapter 2), and people living in these regions appear among those affected earliest (see Chapter 4) and most significantly (see Chapters 4, 7 and 9) by global environmental change.

In this concluding chapter we shall first look at the interlinkage between regional (here: alpine) and global phenomena, we then summarise what we already know about global environmental change in alpine regions, and finally, by looking at those questions of research considered to be most pressing, we indicate possibilities for future analysis.

1 THE GLOBAL EMBEDDING OF REGIONAL PHENOMENA

How do regional phenomena and global change processes interrelate? Using alpine regions as our example, we may structure the answer fourfold: in terms of recognition, impact, adaptation, and mitigation.

Recognition

First, we find that global change phenomena can be recognised earlier in those specific regions that are more exposed to them. Among these are alpine regions. Let us reconsider solar ultraviolet radiation, for example. As a consequence of the degradation of the stratospheric ozone layer, solar ultraviolet-B radiation has increased during the last 15 years over many locations of the earth. UVB differs from UVA radiation because only the short wavelengths of UVB are energetic enough to damage DNA. Only UVB has increased because of ozone destruction, but still much more UVA than UVB reaches the earth's surface. Alpine regions are an earlier indicator of UVB radiation change, because UVB increases with altitude (by approximately 10–20% – depending on the wavelength and the location – with each kilometre altitude). We may either measure UVB itself, or we may base our recognition of global change phenomena on secondary indicators. In the latter case, again alpine peculiarities enhance recognition. For example, aquatic organisms respond to UV radiation. However, water composition determines how much UV radiation is absorbed before it reaches organisms. While pure water does not absorb much UV radiation the most important protective 'substance' is dissolved organic matter. The higher the concentration of dissolved organic matter, the shorter is the penetration depth of UV radiation and the higher the protection of sensitive organisms. There is significantly less dissolved organic matter in the open ocean and in clear water systems such as polar and alpine lakes. In Chapter 2 measurements from alpine lakes prove these to be early indicators of stratospheric ozone depletion.

The effect of global warming, too, is found to be more pronounced in alpine regions, such as the European Alps. From 1895 (the coldest year in the past two centuries) to 1985 the mean air temperature at mountain stations in the Alps has increased by ca. 1°C. Since then, however, we have seen a tremendous increase of about another 1°C in the last 15 years (again see Chapter 2 for the exact measurements). Warming in the Alpine area is thus more pronounced than average global figures indicate. The IPCC Third Assessment Report states a global average temperature increase of 0.6°C ±0.2°C for the last 100 years.

Impact

It is not only certain effects, such as global warming, that are more pronounced in alpine regions, often additionally it is also their 'unit' impact that is more pronounced in regions of this type. The warming stated above may have much more severe consequences at altitudes with annual average air temperatures around 0°C than at sites with higher annual means because it implies drastic changes in the aggregation state of water. Melting of permafrost soils, glacial retreat, disappearance of permanent snow fields, duration and thickness of ice cover, stratification and mixing of the water body, flushing rates etc. will all be influenced by a shift of 1–2°C. For example, an increase by 3°C means an upward shift of the mean altitudinal snow line by 300–500 m (see Chapter 4).

Beyond the impact on the Alpine ecosystems themselves, we may use the example of change in the Alpine water cycle to indicate the significant impact that this triggers for other regions. The Alpine regions, due to their – relative to the surrounding lowlands – significantly higher precipitation rates contribute to a major share of lowland river water flow. For example, the Alpine 32% of the total catchment area of the river Po contributes 56% of lowland flow. But even more important than the absolute quantity is the timing of this flow. A large portion of the precipitation falls as snow in higher Alpine altitudes and is temporarily stored in the snow cover or in glaciers. During the summertime – and thus the lowland dry season – melt-water is released at a time when precipitation is at a minimum, and water demand is at a maximum. The temperature rise induced by global warming and the subsequent upward shift of the snowline may thus significantly shift Alpine water run-off to lowlands from high-demand periods in the summer to periods where there is much better local supply anyway, in the fall, winter and spring.

Adaptation

The demand for adaptation is inherently linked to the type and significance of impacts in the first place. While this volume focuses on adaptation *within* alpine regions (in particular Chapters 4 and 7 to 9), the aforementioned example of trans-regional water cycle impact indicates the significance of other regions' adaptation to changes originally triggered in alpine regions.

Mountainous areas serve as a particularly important recreation area. A rise in the snowline will force the population from outside the alpine areas to either change its type of recreation activity while still continuing to choose the alpine region for recreation, or to adapt by shifting to other recreation locations. There may be various social feedback loops enhancing such developments. In this volume, in Chapter 4, for example, it is pointed out that it is the lower altitude, relatively modest ski resorts in particular which attract people from nearby densely populated areas, and thus allow their children to make their first attempts at skiing. If global warming shifts snow reliability (as defined in Chapter 4) from a current 1 200 m a.s.l. to, let's say, 1 500 m a.s.l, the reduction in ski-resort numbers by some one third would hit those modest resorts hard, exerting a potentially overproportional impact on skiing. The young generation, in particular, could then lose contact with winter sports, and by choosing other forms of recreation this would make recently undertaken investments untenable.

Mitigation

The interregional embedding of mitigation efforts is probably of greatest interest with respect to a truly global environmental phenomenon, such as the greenhouse effect. The problem facing any single region is that of assessing the correct level of abatement given the fact that much of the abatement benefits accrue externally. Only an indicative answer has been given to this question in this volume: The fact that greenhouse impacts are comparatively larger in alpine regions is likely to induce greater investment in mitigation policies in these regions purely on the basis of self-interest. The other questions that arise are – now from a global perspective – how much *can* a particular region contribute to mitigation at all, and how much *should* it contribute according to some reasonable burden-sharing rule. As we have deliberately chosen a regional focus in this volume, a consistent answer can only be given to the first of these questions (see Chapters 10 to 12). In stating this answer, we find that many of the mitigation options do in fact have a significant trans-regional component. In regulating transport activities and their related greenhouse gases, for example, a policy focusing only on the Alpine area itself could mitigate only part of the respective emissions. Phrased differently, any given emission reduction within the Alpine area can be achieved only at substantially higher costs if the policy is not also introduced simultaneously and in a co-ordinated fashion for areas beyond the Alps. E.g. for trans-Alpine transport the Alpine fraction of the total trip is often a rather insignificant portion. Policy targeted at this fraction only is therefore hardly likely to change transport mode or distance (mostly even route) decisions (see Steininger, 2002, for an analysis of this issue).

The second regional question at the international level concerns the normative aspect of mitigation volume. As we do not use a global framework of analysis in this volume, we cannot consistently draw conclusions concerning the share of mitigation that should be accomplished within alpine regions in particular. However, in this final section it is perhaps appropriate to hint possible answers to this question.

In determining the interregional allocation of abatement efforts it is instructive to refer to the standard efficiency criterion. This tells us, that most emission reduction should be undertaken in those regions where it is cheapest. Within alpine regions, there seem to be some – at least in the long run – comparatively cheap greenhouse gas mitigation options available (or the resources for such options), such as biomass for energy supply. Such options may be comparatively cheap oweing to their resulting side effects, as is investigated in Chapter 11.

But efficiency is not the only criterion worthy of consideration. We also need to consider distribution. Within alpine regions, as well as across alpine and non-alpine regions, we observe significant income inequalities. A socially desirable or 'just' burden sharing arrangement should take into account these differential starting positions of the different regions. Simply considering and splitting future emission *reduction benefits* would not be doing full justice to those concerned. In fact it may be necessary to demand that certain alpine areas face less stringent emission limits than others. Such reasoning is also supported by the fact that the non-alpine regions have been responsible for the large bulk of greenhouse gas emissions in the past (see Chapter 10, where this is shown for the sector of agriculture). Thus, argumentation similar to that used in discussing the North-South burden sharing in greenhouse gas mitigation (see, e.g., Eyckmans et al., 2001) may apply to the burden sharing between high-income and low-income regions within both the North and the South, where remote mountainous areas usually belong to the low-income regions.

2 GLOBAL ENVIRONMENTAL CHANGE: CONCLUSIONS FOR ALPINE REGIONS

In order to understand the problem at hand and to be able to develop strategies of humankind to answer the global change challenge, human dimensions have to be taken seriously and included explicitly in the analysis. What is needed is more profound knowledge about how and to what extent global environmental change is induced by human actions and, on the other hand, to better understand what are the consequences of global environmental change for human actions, social values and politico-economic interactions.

Some of the main conclusions drawn and reported within the specific chapters of this volume can be summarised as follows.

Natural Science Research Results

The alpine environment seems to provide an excellent basis for analysis and for gaining valuable insights on global change phenomena.

Lakes The study of alpine or high mountain lakes, for example, represents a good starting point for several reasons: Alpine lakes are present all over the world and may thus serve as a global observation system. Moreover, they respond sensitively to external forces as direct human impacts in catchment areas have been absent over long periods of time. Thus, it is possible to separate the impact of global change processes from direct human impacts.

- Using these lakes, first, we can clearly show that the phenomena of global change are strongly interlinked. Rising acidity in alpine lakes, for example, implies increased fish toxicity, due to increased solubility of several metals. The pH level of alpine lakes, however, is not monocausally due to acid rain (as originally thought) but is also inversely related to temperature. Thus, in this respect, global warming competes directly with acid rain: First, in terms of the simple temperature effect that enhances lake productivity, and second, global warming triggers more precipitation and thus enhances the weathering rate, and as a consequence, alkalinity in alpine lakes.

- Second, as we have stated above, global warming in Alpine environments is higher than respective global figures, as is the increase in (absolute levels of) UVB radiation that reaches the earth's surface (for all of these conclusions see Chapter 2).

Glaciers The impact of global environmental change on glaciers is significant. In Switzerland, for example, the glaciers have lost about one-third of their volume since 1850. With the further temperature rise predicted, the ice mass of Alpine glaciers could be reduced to only 25% of that found in 1850 within the next 30 years, and to only 5% by 2100 (see Chapters 4 and 5).

Snow cover Snow is always close to its melting point, therefore even minor temperature changes may have a dramatic impact on its presence and on its quality. An in-depth understanding of the various processes (precipitation as snow fall, snow accumulation, snow cover metamorphism) was necessary to derive the conclusions that a 3°C temperature rise would raise the altitudinal snow fall limit by 200–400 m, or that the mean altitudinal snow line would rise 300–500 m, with the result that below 1 200 m a.s.l. closed snow cover throughout the winter season would no longer exist. However, for sites at or above 1 500 m a.s.l. significant trends for the seasonal snow cover could not as yet be found (see Chapter 4).

Hazard formation A further warming by 1 degree will cause a spatial reduction of both glaciers and permafrost comparable to that observed within the last ca. 150 years. In terms of hazard formation, the main effects of this loss of ice are that large areas of unstable, loose material will be freshly exposed to erosion processes and that the release of these masses will be at a higher altitude causing higher energy potential and process dynamics. While there is also a tendency of restabilisation due to the fostering of vegetation as a result of increased warming, the reaction of the cryosphere is more rapid, and thus induces a period of more unstable conditions before the system eventually attains stability.

Political, Social Science and Economic Implications

The most prominent impact of climate change in alpine regions stems from the relevance of extreme weather events. Avalanches, landslides, flash floods or soil erosion will result from more frequent wind storms, rain or snowfall. For exposed settlements and traffic lines safety investments or relocation will have to be considered. Further, due to this impact, as well as to the more general impacts of global environmental change, changes in tourism might occur.

Economic analysis Conventional economic analysis falls short in incorporating and portraying all relevant aspects of global environmental change and its human dimensions. Particularly, in concentrating on flows and not on stocks, and neglecting variations in natural capital, it fails to capture all relevant dimensions of sustainability in an adequate way. Alternative capital stocks and their variations have to be considered in a comprehensive framework and the substitutability of human-made, natural, human and social capital stocks has to be examined in order to be able to evaluate and to compare the (total) social and economic impact of global environmental change (see Chapter 3). Some sectors in the economy are forced to come to terms with such matters earlier than others, e.g. the (re-)insurance sector (see Chapter 5).

Values at risk Data from the insurance industry suggest that damage claims from weather-related catastrophes are increasing in Alpine regions (see Chapters 3 and 7). The increase in risk for such damages (which is determined by the three parameters hazard, vulnerability and value at risk) is strongly caused by the increases in values at risk. The changes from a society predominantly determined by agriculture to one of mobility and leisure have been crucial in this increase in values exposed to risk as they involve (i) increasing demand for new infrastructure both in the old settlement centres and in areas previously avoided, (ii) an additional presence of people (tourists) at specific periods, which multiplies the original Alpine population and (iii) an almost complete loss of awareness of natural hazards and the associated risks, and, as a consequence, a decrease in risk acceptance (see Chapters 6 and 7).

Motivation and competence to deal with risk Although the analysis of adaptation and adjustment processes to global change is often discouraging, it is by no means without promise. The person's or group's capacity to develop and implement protection and prevention measures is said to be influenced by two factors, i.e. motivation and competence. Even if competence were not restricted, the motivation to implement effective measures is often undermined by short-term economic interests. Consequently, not only the incentives of relevant actors but also the necessary changes in decision-making frameworks have to be identified and focused on in order to redirect natural hazard policies effectively (see Chapters 6 to 9).

Natural hazards management Dealing with natural hazards has a long tradition in most countries in the Alps, with the first official authorities for the protection against natural hazards founded in the second half of the 19th century. While the emphasis was first on technical defence works ('active' measures), the focus since has shifted somewhat to complementary 'passive' protection measures, such as hazard zone maps since the mid 1960s. However, the longer the tradition in hazard zone concepts in individual countries, the more the procedures are dominated by a single discipline, either forest engineers (as in Austria) or geologists (as in France). Only in the most recent and still predominantly scientific approaches is an interdisciplinary approach taken (see Chapter 7).

- It is also the institutional setting of hazards control that deserves close scrutiny when determining its effectiveness. Public bureaucracies, for example, not only follow formal goals (general principles of bureaucracy like legality, efficiency and expediency), but also informal goals, for instance: in order to minimise conflicts they regulate a sector rather in correspondence with than against the interests and values of their main clientele (selective clientelism). The result of negotiation processes may then be that the public agency serves the interests of the private group rather than controlling them (capturing of the institution). A change of structures and processes are unlikely because of the resistance of established actors who benefit from the existing situation.
- An example for required institutional change, as suggested in Chapter 8 of this volume, is the reform of Austrian avalanche committees: they should comprise members from outside the community concerned. At present these committees may easily be biased in their assessment of risk situations because of their members' involvement in local economic interests. Also, the decisions of these committees, in the case of catastrophes, need to be subject to follow-up examination. Another example is the necessary refinement of subsidies for protective measures (they should be significantly lower for those that benefit prevailing economic interests than for those protecting existing settlements) and the raising of financial contributions from beneficiaries (land-owners that benefit from protective measures) should be used to contribute to the cost.
- The Austrian Torrent and Avalanche Control System, for example, has mainly three types of instruments at hand: technical structures, biological measures, and land-use regulation. Technical structures like check-dams have immediate effects, thus having higher political value. Further, politicians are also interested in economic benefits (increased property values) and symbolic benefits (proof that they care about the safety of their municipality). As a consequence, we find that although afforestation would provide long-term enhancement of communal safety, it is not adequately promoted (see Chapter 8).

Tourism In addition to the above factors global environmental change also has a direct impact on tourism. Taking Switzerland as an example, today 85% of ski resorts can be considered snow-reliable. A temperature increase by 3°C would raise the guaranteed snow-reliability level up to 1 800 m a.s.l., resulting in a reduction of continuously skiable resorts to only 44%.

Investment in tourism has already reacted to this tendency. For example, during the last 10 years the number of lift facilities in Switzerland has been reduced by 6%, mainly in lower elevations up to 1 000 m a.s.l. But at the same time lift transport capacity has increased by 3%. This is mainly due to replacement of older facilities by more powerful systems. Totally new constructions are focussed on elevations between 2 000–3 000 m a.s.l. The general shift towards more snow-reliable elevations is clearly noticeable (see Chapter 4).

Mitigating global change Besides measures dealing with impact mitigation, sources of change can also be addressed directly. The sectors most probably contributing to global environmental change in alpine environments are agriculture, energy, transport and tourism. The analysis shows that though the potential for mitigation varies between sectors, there is some direct relevance for the Alpine environment on a global and local scale.

- For greenhouse mitigation potential in Alpine regions, this volume finds that one of the most significant contributions of alpine agriculture could be in changing land-use patterns (and thus the related carbon household). In alpine regions, carbon fixation by both plant biomass and top soil in grassland and forest is roughly double that found in arable land. Yet, even the most extreme scenario of a total transformation of all Alpine arable land into grassland (or forest for that matter) – here quantified for Austria – would only fix Austria's current CO_2 emissions of a single year (see Chapter 10).
- Sustainable energy use is – as shown in Chapter 11 – just one of many goals contributing towards more sustainable development in Alpine regions. The use of renewable energy in turn is just one possible and important (albeit not always the most cost-efficient) approach to tackling the severe and long-term problem of global warming. A case study of the Austrian province of Carinthia has shown that under current conditions and without the imposition of severe additional policies, the widespread dispersal of renewable energy sources alone is unlikely to lead to a stabilisation, or even reduction, of the prevailing GHG emission levels in the Alpine area. The steering of consumption patterns as well as the avoidance of excessive rebound effects have been identified here as two key factors for success.
- Inner-Alpine as well as – increasingly dominating – trans-Alpine transport is a major cause of local and global environmental impact. While both technological and demand side management solutions would be capable in their own right of shaping transport in a

sustainable way, each of these 'pure' paths is shown to be extremely costly (in terms of explicit technology costs or in terms of social adjustment). Combinations of the two, however, imply a range of adjustments such as a shift of rail traffic, or extensive use of fuel cells. They would thus allow for the achievement of even stricter environmental goals such as a reduction of CO_2 transport emissions by 80% by 2030 (relative to 1990) in the Alpine area at low long-term economic cost.

In Alpine regions, due to the greater ecological, social and economic sensitivity of the area, and the important cultural heritage of the Alps, decision makers can in some sense be expected to be more willing to take earlier and more decisive action than in other regions. The protocols developed under the framework of the 'Alpenkonvention' are indeed encouraging in this respect, although some probably still have a long way to go before ratification is completed. Policies taken in the Alpine region may also serve as an important model for other socio-economically and ecologically fragile areas both in the developed and in the developing world.

3 FUTURE RESEARCH PROSPECTS

While, as stated above, the results of research on global environmental change in alpine regions have been significant, they also point out further areas in need of attention. Indeed, some of these have become evident in the present volume and will be briefly commented on below.

- The increase of winter precipitation triggered by the temperature rise in Alpine regions may have contradictory effects. The increase in solid winter precipitation (snow etc.) may raise seasonal storage capacity. The rise in temperatures on the other hand will lead to earlier snow melt. The various interactions are in need of further exploration.
- In Alpine regions, permafrost is widely present above 2 500 m a.s.l. Discontinuous permafrost tends to be highly sensitive. A temperature rise may force permafrost to melt, and potentially lead to a variety of slope instabilities in bedrock and non-consolidated sediments. Mitigating risks due to slope instabilities in Alpine areas requires systematic observation and monitoring of actual critical situations and an improved understanding of the processes at work. Also, technical constructions built on permafrost ground (e.g. avalanche protection barriers) may be exposed to accelerated creeping and settlement, if temperatures should further rise.
- Adaptation, too, may have impacts which today still go largely unrecognised. For example, artificial snow production has been

increasing continuously over the last ten years. Currently 5% of Swiss slopes, 20% of Austrian slopes, and up to 30% in specific ski resorts in Australia, for example, are prepared with artificial snow. The resulting changes in winter snow conditions are expected to have an ecological impact. First, the physical and chemical changes in ski slope snow cover are unknown, as is how these interact with and impact on local vegetation and soil. Further, the use of additives in artificial snow production and/or in ski slope preparation, and their potential impact on fragile Alpine environments is not fully understood. Initial results of research along these lines are reported in Chapter 4; they indicate, for example, that the application of artificial snow seems to favour late flowering species and snow bank communities and to hinder early flowering species and wind-edge communities.

- The strong dependency on snow reliability does not bode well for winter tourism. Alternatives to traditional winter tourism activities have to be developed. This is a long-term goal and might be the only way forward for ski resorts affected by climate change. In addition, this policy might also be consciously directed so as to better protect the ecologically fragile Alpine environment.
- Though data are available from main reinsurance companies about disaster trends and the development of claims burdens in recent years, more accurate estimations of the economic consequences and total socio-economic cost are still outstanding. In order to cope with the increase in the damage potential due to natural hazard events and global change processes in Alpine environments, new strategies have to be developed.
- Identifying the correct combination of different tools for coping with the risks resulting from the limits of prevention measures requires more in-depth-analysis. E.g. what is the best combination of protective structures and secondary structural measures, hazard zoning, warning systems and disaster management (i.e. integrated risk management)? Conventional systems seem to hide remaining risks, instead of addressing and challenging them explicitly.
- Psychological factors, such as social norms, values, knowledge, motivation and competence are well known theoretically to influence human capacity to develop and implement protection and adaptation measures in response to the impacts of global environmental change in alpine regions. Yet, their empirical relevance, and as a consequence, the incorporation of this knowledge in adaptation and protection policies have received remarkably scant attention in the literature.
- On the natural science side, for a better handling of mitigation options in alpine regions the soil emissions of N_20 need to be better understood. The microbacterial transformation processes of nitrogen (nitrification and denitrification) respond to the higher turnover of

nitrogen in agriculture. The multiplicity of factors relevant, however, calls in particular for more long-term analyses in this area.

- On the social science side, a lot of questions in mitigation and response options and on actors' reactions to these remain open. For one type of activity of particular importance in the Alpine region, transport, for example, this involves looking at the individuals and aggregate reaction to the introduction of passenger transport road pricing (volume, route change, mode change, threshold levels), and to concepts of local sustainable transport schemes (such as the introduction of a 'bus day' in remote rural areas, ensuring a dense schedule for that particular day, with the day shifting among regions).

This book explicitly focuses on the social science dimension of global change with the intention of moving beyond a pure natural, pure social or pure economic science analysis. Such an overlapping approach is particularly important given the prominence of global change processes and their human dimensions in alpine environments and given the prevailing dominance of natural science-based projects in global change research. The challenge is to combine the questions, to interact during analysis and to collate findings in such a way that interdisciplinary work becomes both fruitful and adequate for the real questions and problems at hand. This book attempts to take us further in this direction.

BIBLIOGRAPHY

Eyckmans, J., V. van Steenberghe and D. van Regemorter (2001), Is Kyoto Fatally Flawed? An analysis with MacGEM. FEEM Discussion paper 43.2002. Milan: Fondazione Eni Enrico Mattei.

Steininger, K.W. (2002) Environmentally Counterproductive Support Measures in Transport: A Transboundary Pollutant? In: Rauscher, M. (ed.), *Environmental Policy in an International Perspective*, Kluwer (forthcoming).

Index

Note: Page references in bold type indicate the main text entries.